OCEAN

DAVID ATTENBOROUGH

OCEAN

EARTH'S LAST WILDERNESS

COLIN BUTFIELD

JOHN MURRAY

First published in Great Britain in 2025 by John Murray (Publishers)

1

Text copyright © David Attenborough Productions Limited
& Colin Butfield 2025

Illustrations © Jennifer N. R. Smith 2025
Southern Ocean chapter illustration based on the
Gott-Goldberg-Vanderbei projection courtesy of Robert Vanderbei

Text design by Janette Revill

A CIP catalogue record for this title is available from the British Library

Hardback ISBN 978-1-399-81850-6
Exclusive hardback ISBN 978-1-399-82762-1
Trade Paperback ISBN 978-1-399-81851-3
ebook ISBN 978-1-399-81853-7

Typeset in Warnock Pro by
Palimpsest Book Production Ltd, Falkirk, Stirlingshire

Printed and bound in Great Britain by Clays Ltd, Elcograf S.p.A.

John Murray policy is to use papers that are natural, renewable
and recyclable products and made from wood grown in sustainable
forests. The logging and manufacturing processes are expected to
conform to the environmental regulations of the country of origin.

Carmelite House
50 Victoria Embankment
London EC4Y 0DZ

www.johnmurraypress.co.uk

John Murray Press, part of Hodder & Stoughton Limited
An Hachette UK company

The authorised representative in the EEA is Hachette Ireland,
8 Castlecourt Centre, Dublin 15, D15 XTP3, Ireland
(email: info@hbgi.ie)

CONTENTS

CONTENTS

PART THREE

AUTHORS' NOTE

In writing this book we have drawn upon, and been inspired by, the work of hundreds of scientists from a vast range of disciplines. The key sources we have used are listed in a section at the end of this book, but there are also a few great scientists who were kind and generous enough to give their time and expertise to ensure *Ocean* is as up to date and accurate as possible.

We would particularly like to acknowledge Dr Mark Belchier, Government of South Georgia & the South Sandwich Islands and British Antarctic Survey; Madi Bowden-Parry, University of Exeter; Rod Downie, Chief Polar Advisor, WWF; Nico Koedam, professor emeritus (Vrije Universiteit Brussel), presently affiliated to Universiteit Gent, Marine Biology Research Group; Dr Tom Bech Letessier, University of Plymouth and University of Western Australia; Professor Daniel Mayor, University of Exeter; Professor Michael Meredith, British Antarctic Survey; Professor Pippa Moore, Newcastle University; Professor Callum Roberts, University of Exeter; and Dr Dan Smale, Marine Biological Association of the UK.

A special debt of gratitude is due to Dr Casper van de Geer of the University of Exeter. Casper has worked closely with us throughout the writing of this book and his excellent research and wide knowledge is ever present on these pages.

Thank you all.

David Attenborough and Colin Butfield
February 2025

PREFACE

My earliest memory of the ocean is of a tropical lagoon.

Ammonites rose and fell in the warm water column, occasionally propelling themselves forwards, their curled ram's horn shells surprisingly streamlined in the sediment-rich water. Bullet-shaped belemnites squirted ink as they fled from a predator above the oyster beds. I was sure there must be hundreds more species living in this rich sea, but I had yet to glimpse them. I resolved to keep looking.

This tropical lagoon was in fact in my imagination, fired by exploring an old limestone quarry in Leicestershire, some sixty miles from the coast. For a small boy in the 1930s this was a marvellous place for adventures, and the knowledge that millions of years ago it would have been a warm and wild lagoon only increased its appeal. Here I could spend days searching for treasure buried in rocks laid down in ancient tropical seas. Holding the fossils of long-dead sea creatures that I had chipped out of the rock, knowing my eyes were the first ever to see them, ignited my curiosity. I would spend much of the rest of my life wondering what lived below the surface of the ocean.

Palaeontologists realise that the ancient ocean they assemble from rare, preserved fragments is incomplete. Many of the wonders that swam in the seas of the Jurassic or Cretaceous periods may simply be unknowable – if they were never fossilised then they cannot be discovered. There are some parallels with the quest to understand the ocean of my lifetime.

Throughout my first few decades, we too only caught glimpses of life in our ocean. There was so much we had yet to understand. We were finding creatures at great depths but knew little more of how they survived, or with which other species they shared the deep, than we knew of the seas in which my ammonite fossils once swam. We were hunting whales in vast numbers but knew next to nothing of the effect that might have on the wider ocean ecosystem. We could take pictures of coral reefs but could not explain why they harboured such a wonderful array of diversity.

I have been fortunate enough to live for nearly a hundred years. During this time we have discovered more about our ocean than in any other span of human history. Marine science has revealed natural wonders a young boy in the 1930s could never have imagined. New technology has allowed us to film wildlife behaviour I could only have dreamed of recording in the early stages of my career, and we have changed the ocean so profoundly that the next hundred years could either witness a mass extinction of ocean life or a spectacular recovery.

I will not see how that story ends, but, after a lifetime of exploring our planet, I remain convinced that the more people enjoy and understand the natural world, the greater our hope of saving both it and ourselves becomes. When Colin and I decided to write this book, we wanted to convey the thrill of discovery, be clear-sighted about the danger our ocean is in and, above all, share the stories we feel could inspire a new generation to look beyond the shore and beneath the waves.

David Attenborough
February 2025

PART ONE

IN THE LIFETIME OF A SINGLE BLUE WHALE

E verything we knew came from the dead; specimen jars of faded life that had been dragged from unimagined depths, tales passed down from explorers and fishers, remnants washed up on beaches or stranded on rocky shores. A hundred years ago much of our ocean was a mystery – a vast, hidden world visible only in the imagination.

At the time, our knowledge of life on land was well advanced. Yet of the species occupying the other two-thirds of our planet's surface and 99 per cent of its habitable area, only fragments were known; ephemeral glimpses encouraging us to seek further and deeper. Little made sense at first – dazzling diversity flourished in nutrient-poor waters, remote underwater mountains far from the nearest continent were thronged with life, and every now and again animal remains were found that defied explanation. But gradually we found clues that led to ideas, ideas led to hypotheses, and hypotheses became revelations. Advances in technology meant we could watch, track and map ocean life. Little by little the ocean gave up some of its secrets.

OCEAN

PACIFIC OCEAN, 1941

Two hundred kilometres off the coast of California a blue whale surfaces within sight of a convoy of heavily armoured grey ships heading out to sea. A sailor on board, who grew up fishing these Pacific waters, recognises the distinctive spout. On the top of its head, the blue whale has powerful muscles in the form of a V framing its nostrils – the twin blowholes common to all baleen whales. When the muscles relax the blowholes are plugged to stop water entering, but when the mighty whale surfaces the muscles contract and the blowholes open, allowing it to breathe. Easily distinguished from the flat, low cloud produced by the humpbacks that frequent this area, the blue's spout soars high into the air at a speed of nearly 600 kilometres an hour, powered by the immense lungs of the largest animal on the planet.

It is an eight-year-old female. She has been feeding in the cold, rich waters of Alaska and is now travelling thousands of kilometres south, past towering kelp forests and nutrient-rich river deltas. She pauses to rest and feed on the abundant life gathered at underwater mountains known as seamounts, and at upwellings where nutrient-rich water attracts life from the open ocean. Soon, she will veer back towards land where the coastline's vegetation changes from fir tree to cactus, in search of warmer, sheltered waters to give birth to her first calf.

Unbeknown to both whale and the sailors, as the human world endures its most turbulent time in modern history, swathes of the ocean are about to enter a period of relative calm. Despite atomic tests and naval battles, the horrors of the Second World War will bring respite to certain parts of the ocean. Areas including parts of the North Sea in Europe will become far too dangerous to search for fish and will experience a dramatic recovery in marine life. An accidental experiment, for the grimmest of reasons, it will nonetheless yield the first large-scale evidence that the ocean can recover faster than we had ever imagined.

LONG BEACH CALIFORNIA, 2024

Now nearing the end of her life, our whale is making this journey one last time. She surfaces 100 metres from a small boat full of tourists. The biologist on board tells of whales' feeding and migration habits while the whale watchers hope for a shot of a fluke or a spout. The loading cranes of the port of Long Beach have only just slipped out of sight as the boat reaches the deeper water where blues, and their close relatives fin whales, can be sighted traversing the route between the Channel Islands National Park and the surf beaches and shipping industry of the Los Angeles coast.

Her life has been a fortunate one. She was among the very few of her kind to survive commercial whaling, and she has travelled the length of this coast dozens of times, covering tens of thousands of kilometres. It is likely that she produced a calf every two or three years of her long life, during which time her species has gone from the brink of extinction to the beginning of recovery.

Over the course of her lifetime we have made a journey of our own: from viewing whales as a source of oil to recognising them as a source of wonder and even kinship, their deep intelligence and complex social interactions resonating with our own. Wise and passionate people ended the era of industrial whaling and we embraced a new relationship born of scientific understanding, foresight and compassion. Despite that journey, we have yet to extend the same foresight to protecting their home. For many of us the world beyond the beach is dark, threatening, other-worldly – out of sight and most certainly out of mind. But gradually all that is changing. Decades of dedicated scientific study, technological advances and a rekindled respect for traditional local knowledge have yielded remarkable discoveries, revealing the central role of the ocean in all our lives and showing us exactly what we must do to restore it to health.

The blue whale surfaces one last time. She dives deep. The whale watchers will not see her again.

The lifetime of a blue whale – some ninety years – is a handy benchmark for our journey of modern ocean discovery. Today's perspective on the blue whale's world is unrecognisable from the way we viewed the ocean when our whale was born in the 1930s.

While seafaring cultures, most notably that of the Polynesians, had thousands of years of knowledge about *how* to expertly traverse the ocean, and commercial fishing nations such as the UK and USA had learnt *how* to efficiently target huge stocks of fish, science had yet to explain *why* ocean currents behave in the way they do or why particular fish are found in certain places at certain times. To understand those questions, we needed new ways to look at our planet.

'Ocean' is a far more appropriate name for our world than 'Earth'. Today just over 70 per cent of our planet's surface is covered in salt water connected into a single planetary ocean. The shifting of tectonic plates and the ebbs and flows of ice ages govern the map of our ocean, but for the last 10,000 years or so its major points of connection have been as they are today. When our blue whale was born we could only view those connections from the perspective of the surface and the land. We knew the shapes of the continents that framed the ocean, we had mapped the gaps in land that connected the Red Sea to the Indian Ocean, the Mediterranean to the Atlantic and the Atlantic to the Arctic. But the ocean only really makes sense in three dimensions, so in order to truly understand it we needed a whale's-eye view of the world.

It was the advances in sonar during the Second World War that began to make this possible, and by the time our whale was a teenager we were starting to get our first real view of the seabed. The data provided by sonar revealed that the ocean floor wasn't a flat, featureless plain as many had thought but instead contained vast mountainous ridges, deep trenches and volcanoes. It had features and regions as clearly defined as any on land. We began to think of the ocean in terms of five major connected basins: Arctic, Atlantic, Indian, Pacific and Southern – though the Southern Ocean wasn't officially recognised as a distinct ocean basin until 2021.

The Pacific is by far the largest of the five, covering almost half the ocean and big enough to accommodate all the land on Earth. It was so named by sixteenth-century Portuguese explorer Ferdinand Magellan because of the calmness of the waters he encountered. This may sound unlikely to anyone familiar with the Pacific winters off Hawai'i or northern California, but Magellan had entered it via the lethally treacherous strait at the tip of South America that still bears his name, so, by comparison, a balmy day in the Pacific might well have felt peaceful. The Pacific is so vast that you could leave Melbourne, Australia, and reach either the southern end of Chile or approach the Arctic via the Bering Sea without ever leaving it.

While all five ocean basins are connected, it is how and where they connect that is important to our understanding of the way currents, nutrients and wildlife move around the ocean. The route from the Pacific to the smallest of the basins, the Arctic, is through the narrow, shallow Bering Strait. Relatively little water or wildlife travels through this gap. By contrast, at the point where the Pacific hits the youngest of the world's ocean basins, it enters the marine equivalent of a blender – the Southern Ocean. Waters from the Pacific, Atlantic and Indian ocean basins all connect with the Southern Ocean and are mixed up by the Antarctic Circumpolar Current that powers its way clockwise around the entire continent of Antarctica.

By the mid-1950s the main ocean basins, their sub-divisions (Baltic Sea, North Sea, Gulf of Mexico and others) and their points of connection were common scientific and political language. This is significant because although we had long known that different ocean habitats flourish in different parts of the world – coral reefs in the tropics, kelp in more temperate water – we now began to view the seawater itself as part of one system, our many seas as just one single ocean.

Indeed, the more we looked, the clearer the evidence appeared that certain species could be found throughout the ocean: the blue whale, for example, had been recorded across all the ocean basins; only the frozen parts of the Arctic and Southern Ocean were out of their reach, something that will surely change over the coming years as whale numbers recover and the sea ice retreats.

By the 1950s our whale was a fully grown adult over 25 metres in length and weighing in excess of 150 tonnes. She was not merely 'big', she was a member of the largest animal species on the planet – far more massive than most dinosaurs. We knew why her species could grow so large – the buoyancy of the seawater enables ocean-dwelling animals to achieve a mass that bone could never support on land – but we did not yet know enough about a blue whale's life to understand why there was an evolutionary benefit in reaching such a size. It would take much of our whale's lifetime to discover that.

A key clue came from a deeper revelation about ocean currents. We had long understood that there were dominant currents at the surface of the ocean, but it wasn't until the 1960s that decades of research by scores of scientists across the world was drawn together to describe a global system of currents known as thermohaline circulation – or the global conveyor belt, as it has now become known.

Named after the two factors influencing the density of seawater, temperature (thermos) and salinity (haline), this system begins with the freezing of seawater in the far north and south of our planet. As the ocean freezes in the Arctic, it leaves behind its salt, which cannot freeze, making the remaining surface water saltier and denser. This cold, dense water sinks and other surface water is drawn in to replace it, creating a current. The dense water sinking pushes the existing deep water south, and over hundreds of years this slow, deep current moves through the ocean basins to Antarctica where it is joined by more cold, salty, sinking water. The Antarctic Circumpolar Current moves the water clockwise around Antarctica until it is driven north again in two currents – one heading into the Indian Ocean and the other to the Pacific. As they travel north the currents gradually warm and rise towards the surface. The warmed waters continue circulating around the globe, eventually returning to the North Atlantic and up to the Arctic where the cycle begins once more.

The currents bring nutrients from the depths to the surface, enabling the growth of plankton and thus powering almost the entire ocean food web. Scientists have also discovered that by transporting heat from the equator towards the poles, and vice versa, these currents have a profound effect on the world's climate. For example, it is this system that brings warm water through the North Atlantic, keeping the countries of north-western Europe, such as the UK, much warmer than other places at the same latitude.

Ocean currents both local and global are vital to all life

on Earth, not just the blue whale. But the upwelling of currents is thought to have had a particular impact on blue whale evolution because nutrients brought to the surface by upwellings feed the prey of blue whales, and it is their selection of prey that in turn explains why the blue whale is so large. The blue whale sustains its vast bulk by eating immense quantities of some of the smallest animals in the ocean, but there is clearly no need for epic size in order to overpower this prey. Over the course of its ninety-year life, our blue whale will have eaten billions of krill – small shrimp-like crustaceans – and the manner with which it does so requires a quite extraordinary physical transformation.

For all its great size, one of the most noticeable features of a blue whale is its streamlining. Like a slow-motion torpedo, it displaces water around its head and along its flanks as it effortlessly slides along at a cruising speed of around 10 kilometres per hour. But when it finds a swarm of krill it can open its massive jaws to nearly 90 degrees by dislocating the lower mandible and expanding the pleats of skin beneath its mouth, enabling it to swallow 80,000 litres of krill-filled water in a single gulp. This it filters through baleen plates that hang on either side of its upper jaw and allow the water to spill out while trapping the krill inside. By comparison we humans can manage a paltry 0.07 litres, sans krill. When feeding, a blue whale's mouth can balloon to the equivalent size of the rest of its body, giving it the silhouette of an enormous tadpole.

This is known as lunge-feeding, and as far as we can tell from the fossil and geological records, it appears to have evolved in baleen whales a little over 7 million years ago, around the same time, scientists now believe, as a notable increase in nutrient-rich ocean upwellings. These upwellings would have stimulated blooms in plankton, small fish and crustaceans that could be exploited by those species able to lunge-feed and gulp large quantities quickly. Unlike other baleen whales, blue whales evolved to be specialist hunters, with krill making up nearly their entire diet. And this specialisation further influenced their appearance and behaviour.

There are at least eighty-five species of krill found throughout the ocean, and all of them feed on plankton. The term 'plankton' comes from the Greek word for 'drifter' and is used to define any microscopic animal (zooplankton) or plant (phytoplankton) that drifts with the ocean's currents. Krill swarm in huge numbers, but those swarms aren't constant and can be significant distances apart. The ultimate krill predator therefore needs to be able to travel efficiently for many weeks at a time without eating, but when it does get the opportunity to feed it must be able to consume it in immense quantities. A huge whale – with an equally huge mouth – that is streamlined yet able to live off its fat stores for months while it slowly cruises vast distances is a highly efficient solution for surviving on krill.

But something even longer than our blue whale also hunts krill. Nearly 200 metres in length, it hovers in the water column at night. Its ghostly white colour glows as it emits bright blue bioluminescent light to attract prey towards its curtain of stinging tentacles. It might look like a single creature, perhaps a long, thin jellyfish, but this is a giant siphonophore – a colony of genetically identical individuals known as zooids, each with a special function in supporting the colony: some catch prey, others digest it; some swim, others reproduce. The siphonophore is hovering here at night because this is when vast numbers of animals rise from the depths to feed: the diel vertical migration – the largest daily movement of biomass on Earth.

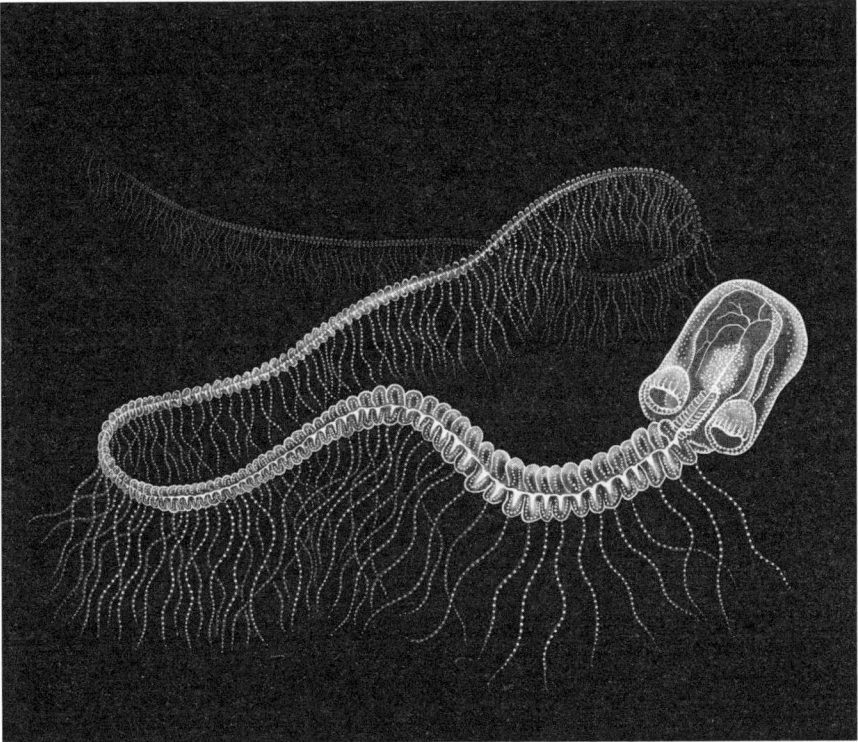

Growing longer than a blue whale, the giant siphonophore is a colony of individual organisms known as zooids.

While it had been known for many years that there was some vertical migration in the ocean, it wasn't until around the time our whale had reached middle age that scientists had the technology and research capability to uncover the sheer scale of the vertical migration happening each day across the ocean. Until this point most of us had thought of marine migrations as travelling horizontally to new feeding or breeding areas, as many whales, tuna or seabirds do. But the research revealed that vast numbers of krill, lanternfish, squid and countless other species were descending beyond the reach of light during the day to avoid visual predators. Once night fell, they would seek phytoplankton-rich waters and head towards the surface. Like all other migrations, this too had its predators. Some travelled with the vertical migration, picking the travellers off as they rose, while others, such as the giant siphonophore, lay in wait, their vast trap set across the equivalent length of two consecutive football pitches. With billions of animals taking part in this migration each night, there was plenty for the predators.

The second half of our whale's life mirrored a time of incredible discoveries. We were beginning to see the ocean in three dimensions, and as our technology developed the discoveries kept on coming. Technological leaps were deployed by marine scientists to give a completely new perspective on the deep ocean. Underwater submersibles revolutionised ocean exploration, and by the 1970s they had been used to discover a totally new form of life around the deep-sea vents. And as the technology improved further, it revealed new wonders. Remote vehicles could now spend days on the seabed exploring, all the while sending images and data back to the surface. Nowadays anyone with an internet connection can watch live-stream video from remote-vehicle missions thousands of metres below the surface.

Sending new technology to the deep transformed our perspective on the ocean, but so too did sending it to space. In 1957 we launched our first satellite. Over the next sixty years satellite technology advanced to such an extent that it could be used to discover thousands of seamounts previously hidden below the ocean surface, track tagged animals to reveal their migrations and provide insights into ocean hot spots and superhighways for marine wildlife. A combination of visual sightings and tagging built up a picture of blue whales across the world. Distinct subspecies and populations were identified and some truly massive migrations recorded that revealed hitherto unknown connections between different parts of our ocean: for example, research showed that our whale, a subspecies called the northern blue whale, would probably spend her entire life in the north-east Pacific travelling from as far north as Alaska to as far south as Costa Rica.

But satellites and submersibles could only provide part of the picture. To understand the connections between features in the ocean and species migration – why certain species take the routes they do – requires detailed mapping, and for this sonar has once again been transformational.

Since the late 1980s multibeam sonar has been providing highly detailed images of a large area of the sea floor. The most recent versions send out over 1,500 sonar soundings a second from a ship. The sonar beams fan across the bottom creating a sound map that a computer can turn into a visual representation of the sea floor. In fact, by the end of 2023, 25 per cent of the entire sea floor had already been mapped to a resolution of 100 metres or more, and a major project is currently under way to map the entire seabed by 2030. The picture is complemented by split-beam sonar, which sends and receives sound pulses in the water column and as a consequence can detect and create sound maps of the species that are swimming in that area, although of course it can only detect what is there at that exact moment.

Even that limitation can now be addressed by advances in DNA sampling, which enables scientists to detect what has passed through a water column. Environmental DNA tests sample the water for traces of marine species – skin, faeces, mucus – analysing the DNA of the species that have swum through an area in the preceding few hours without ever needing to catch or even see them.

These astounding technological advances, combined with ships that can stay at sea for months at a time and remote monitoring systems in buoys that provide constant year-round information on waves, water temperature and chemistry, have revolutionised our understanding of the ocean. In the lifetime of a single blue whale we have gone from skimming the surface of our ocean to deeply comprehending its importance. Yet while these new technologies have given us a whale's-eye view of the ocean over the last ninety years, they have also enabled us to change that whale's world beyond recognition.

It is little surprise that when we overlay a map of global fishing effort (where we have fished the most) on a world map it correlates with where science has recorded the greatest concentrations of nutrients and most significant gatherings of marine life. Rather like whales, fishing boats can also now travel for months, detect seamounts and use sonar to locate their prey. We have become so good at catching fish that, it has been calculated, as of 2024 humans have reduced the biomass – the *life* – in the ocean by 2.7 gigatonnes. For context, the entire human population consists of about 0.4 gigatonnes of biomass, so one can imagine the imbalance created by removing almost seven times that amount of life from the ocean ecosystem.

But the changes in our whale's world are not limited to overfishing. The blue whale, like many other open-ocean species, has changed its diet, behaviour and navigational ability over millions of years to efficiently exploit its place in our planet's diverse and complex ocean – and, as we have seen, this niche is particularly narrow in that the whale predominantly feeds on krill and its body is perfectly evolved to collect this source of food. This makes it a highly efficient predator but also highly susceptible to anything that changes the availability of krill or its ability to find it.

We still don't know exactly how blue whales blend their different senses to traverse the ocean, but we do know that sound is extremely important and that there are broad routes along which they migrate at different times of year. Further, we will have affected their senses and those routes in more ways than we can possibly comprehend.

Tens of thousands of large ships transporting goods

around the world generate noise and accidentally but inevitably strike migrating whales. The warming, acidifying ocean is changing the distribution of life within it, making time-worn feeding and breeding patterns unreliable. By deliberately removing the larger species in the ocean for food, our nets, trawls and dredges often also destroy or damage entire habitats, disrupting intricate food webs in ways we have yet to fully discover.

But with our newfound understanding we have also learnt about the ocean's power of regeneration. We now understand so much more about where life flourishes and how we can help it to do so. We have recorded examples of restoration and recovery, and we can – if we choose – monitor and alter our fishing practices to achieve a balance where the ocean can both provide for our needs and thrive. If a calf were born to our blue whale today then it could very well live into the twenty-second century. In a world where we apply the same foresight and understanding that once saved her species to protecting her home, she could live to see a wondrous transformation. Her feeding grounds in the cold, high-latitude seas will be full of plankton, krill and countless species of fish. Her calves will be born in safe waters fringed by mangroves and corals. When she crosses the open ocean her migration routes will be free of nets and the seamounts she passes will be full of life. And when she comes close to shore, the seabed will be alive once again with kelp, corals, mussels, lobsters and oysters. Perhaps on her voyages she will also pass our descendants – members of a society in balance with the natural world that provides it with food, livelihoods and inspiration, living in a time in which humankind has grown beyond trying to rule the waves and instead has finally succeeded in thriving alongside the greatest wilderness on Earth.

Over the lifetime of a single blue whale we have discovered more about our ocean than in the rest of human history combined. But will we discover the foresight to help the ocean recover from the damage we have inflicted upon it during the same period? To answer that question we must leave our whale behind and dive into each of our ocean's most important habitats. From the vast open ocean to mysterious deep-water vents, the frozen sea of the Arctic to the wild Southern Ocean, and isolated seamounts to dense underwater forests, dazzling coral gardens and tangled mangroves; all contain their own evolutionary wonders, inspire their own human guardians and reveal clues to the future of life on Earth.

PART TWO

OUR OCEAN WORLD

1

CORAL REEF

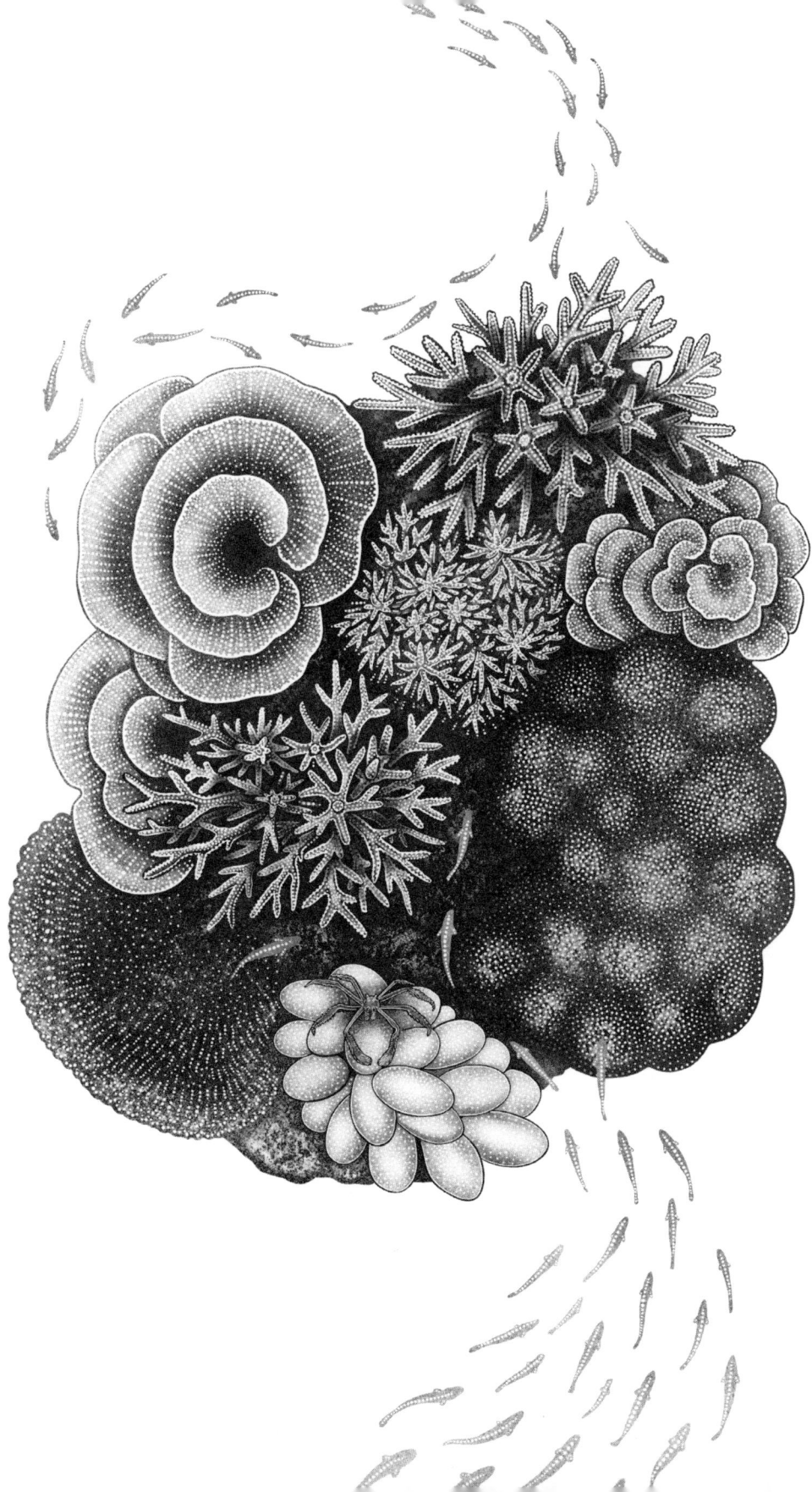

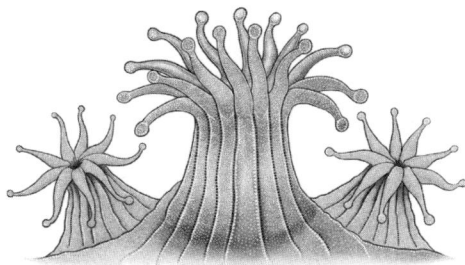

CLOWNFISH, GREAT BARRIER REEF, AUSTRALIA

In the tropics, constant sunlight keeps coastal waters warm year-round, and as a consequence, this is where we find the most dazzling array of life anywhere in the world's ocean.

The first time I used scuba gear, to dive on a coral reef in 1957, I was so taken aback by the spectacle before me that I momentarily forgot to breathe. Nothing I had ever seen on land came close to the sensory overload of so much life, of such diversity, right before my eyes. I could have spent days swimming above it and never tire of the colours, the movement, the interactions. It is life at its most mesmerising.

The dive was on a shallow, warm water coral cay on Australia's Great Barrier Reef. The corals that thrive in these conditions require reliable sunlight and warm water to survive and therefore are typically found in the shallow waters of tropical seas. As a result, the most spectacular corals are often within a few metres of the surface. You don't need scuba gear to get down to them – snorkelling and duck-diving would be sufficient. But scuba allows you to stay in the same place for long enough to start noticing the details of reef life – the countless tiny fish swimming between coral branches, the barracuda waiting under a rock shelf for a flash of silvery prey to pass, the lionfish with its fanlike fins hovering near a rock and at a passing glance easily mistaken for a plant.

Back then the Great Barrier Reef was largely intact. I'm sure there must have been some minor fishing or pollution damage somewhere along it, but at the time it certainly seemed untouched to me. It was common to hear from Australian naturalists that one could snorkel or dive on almost any one of the 3,000 individual reefs along its 2,000-kilometre length and witness one of the greatest natural sights on Earth.

We have known that coral reefs harbour astonishing diversity ever since we first set eyes on them, and I knew of early Hawaiian and Polynesian oral stories passed down over thousands of years that spoke of the importance of coral reefs and the rich life among them. I had also read more recent accounts from explorers and scientists, but words cannot prepare you for actually seeing so many different species, all with their own way of overcoming life's trials, somehow fitting together in an ecosystem so vivid and vibrant. I had never seen anything on land to rival it. Even though we know that a tropical rainforest harbours extraordinary animal diversity, you see relatively little of it on a single walk. Yet on that dive, which lasted perhaps thirty minutes or so, I saw more species of animals than I could have begun to count, let alone identify.

Back then, the aqualung was a relatively recent invention and detailed reef studies were limited. Many of the interactions between species or food chains on a coral reef that we now think of as common marine biology knowledge were unknown at the time. On that dive, one species in particular caught my eye. Weaving in and out of the tentacles of an anemone was a small fish no longer than my index finger. It was bright orange with three white bands around its body and black lines fringing it. Today children across the world would instantly recognise it as Nemo – a clownfish – but at that time I had neither seen nor heard of one. My first thought was to wonder why it was not being stung by the tentacles of the anemone. When I saw more clownfish behaving similarly in other anemones nearby, I realised there must be a special relationship between fish and anemone. But what sort that was, and how the relationship had evolved, I had no idea.

Since then, scientists have discovered that out of some 1,000 anemone species only ten have evolved this relationship with clownfish. But for those that have, it is central to their lives. Clownfish have even evolved an especially thick layer of mucus on their skin to protect against accidental stings, but they never try to nibble the nutritious tentacles of the anemone and live surrounded by them. In return for the protection the anemone provides, the clownfish clean the anemone and chase off predators. The anemone also benefits from nutrients excreted by the fish and their symbiotic algae grow faster and produce more food than in anemones that lack fish.

It has taken the whole of my lifetime for science to begin

to explain how a reef community works and why its immense diversity matters to all of us. Decades of research has revealed that, despite occupying just a tenth of 1 per cent of the sea, coral reefs support a third of all marine species. Even so, we have only just begun to uncover their secrets. So far, we have found over 4,000 species of fish on coral reefs, but it is thought that hundreds of thousands of undiscovered animals and plants live there too.

Indeed, it is only in the last couple of years that we have understood that it is the very diversity of the reef community that attracts the clownfish to a reef in the first place. Researchers have discovered that young clownfish out in the open ocean don't just swim to the nearest reef once they are big enough to do so. Neither do they return to the reef of their parents. Instead, they select a reef by listening. They detect very distinct sounds – the clamour of shrimp and clams and the drumming of swim bladders. The noisier the reef the better, because this means it is more full of life, and the more diverse and populous a reef is, the more resilient it is to the numerous threats that they face today.

On my first coral reef dive it was colour and movement that held my attention, but were I able to return today I would also be searching for sound, trying to decipher the static white noise into pops, whistles and clicks and hoping they would be of sufficient volume to attract more life from the open ocean beyond.

DA

Imagine you have never heard of a coral reef and one day find yourself presented with an aerial photograph of a typical coral atoll. The ring-shaped reefs, with dark blue ocean encasing each azure lagoon, might suggest there could be some upwelling of nutrients from the deep. But the small, low-lying islands the reefs encircle have no rivers carrying nitrogen and phosphorus from great rocky mountain ranges to fertilise the ocean, no deltas with rich sediment, no continental shelf. Your rational conclusion would therefore be that these are nutrient-poor waters, and that in all likelihood there should be few species living there. You would have little reason to suspect that this is the most biodiverse habitat in the ocean.

Yet in the warm seas around the equator, between the tropics of Capricorn and Cancer, thousands of square kilometres of coral reef flourish, as does the life within them. The reefs take many forms. Some are relatively small and isolated, like those surrounding the countless islands of the South Pacific and Indian Ocean; others, such as Australia's Great Barrier Reef or the Mesoamerican Reef of the Caribbean, are vast structures more than 1,000 kilometres long.

It's important to distinguish between 'coral polyp', 'coral

colony' and 'coral reef', as they are all commonly, and confusingly, referred to simply as 'coral'. The reef is a limestone structure often built over thousands of years as each successive generation of hard coral grows on top of the skeletons of the last. There are many different species of coral, but they are typically grouped into 'hard' reef-building corals and 'soft' corals that will still grow on reefs but do not leave skeletons behind. The live part of the reef is the area on top where coral colonies grow, each of which is made up of coral polyps – sometimes vast numbers of them.

Coral polyps are animals that are quite closely related to jellyfish. A single square metre of a healthy coral colony may have more than 10,000 polyps. Each individual coral polyp sits upside down in a tiny cup called a corallite that it makes from limestone, its tentacles pointing out into the water. It uses these tentacles to catch plankton. But in these nutrient-poor waters the plankton it harvests would not, alone, be nearly enough to sustain the coral polyp.

How, then, do these seemingly unproductive waters host the most dazzling display of life anywhere in our ocean? Charles Darwin wondered the same thing, and the puzzle became known as Darwin's Paradox. He solved part of it in 1842 in a now famous book, *The Structure and Distribution of Coral Reefs*, in which he identified the need for shallow waters for most types of coral to grow. But he could not at that time explain how so much life could flourish in waters that should be low in nutrients. It would take a further 200 years to find the answer, but we now know the secret of coral reefs, and it is truly wonderful.

At the heart of the answer is one of the most remarkable and successful mutual relationships in the entire natural world, that of the coral polyp and a type of single-celled alga called a zooxanthella. Zooxanthellae live in the tissues of coral polyps and have a very special relationship with their host. The coral polyp provides the zooxanthellae with shelter as well as the carbon dioxide and water needed to convert the sun's energy into food via photosynthesis. In this process the zooxanthellae take the abundant sunlight and transform it into the food the coral polyps need but the surrounding waters cannot provide. The polyps then use the sugars, fats and oxygen produced by the zooxanthellae to grow and respire – the process of turning glucose into energy – which in turn provides the carbon dioxide and water required by the zooxanthellae.

It is true closed-loop manufacturing. Essentially, the zooxanthellae and coral polyps are passing food to one another in a highly efficient way; in fact as much as 90 per cent of the organic material produced by the zooxanthellae is used by the coral polyps. They are assisted by a whole community of microbes living in and on the polyps that provide nitrogen, remove ammonia and perform a host of other beneficial functions. On land, a typical plant would require many external relationships – with insects, fungi and other living things – to achieve the same ends.

It is in this way that coral reefs manage to grow in nutrient-poor waters. Indeed, the low-nutrient water is vital; algae growth in the water column needs to remain limited in order for the water to stay clear enough for the zoo-xanthellae's photosynthesis to work efficiently. These remarkable relationships have transformed what should have been a barrier to life into the base of a habitat so rich that its biodiversity is rivalled only by a pristine tropical rainforest.

The coral skeleton itself is ghostly white, much like chalk, but the zooxanthellae create protein pigments of the varied colours we associate with coral reefs. We don't yet know why all the different colours are formed, but studies have shown that some of them create a kind of sunscreen, protecting the coral from harmful ultraviolet rays. Scientists also think that the mesmerising fluorescent colours of some corals may serve an evolutionary function by filtering different wavelengths of light for the zooxanthellae so that their photosynthesis is more efficient. They do this by absorbing light in one colour – usually blue – and emitting it in another – typically red. Blue light and red light are well known to be the most effective for photosynthesis, and zooxanthellae require both. However, while the longer wavelengths of blue light travel well through water and are therefore abundant in the ocean, the same is not the case for shorter wavelengths of red light. Scientists believe that because the fluorescent proteins take in blue light and emit red light, the wonderful glowing colour may be created to help the zooxanthellae grow!

Coral polyps are packed tightly and their tissue forms a continuous layer that covers the entire colony like a skin. As the colony grows and matures, individual polyps specialise. In sunnier parts of the reef the polyps are tightly packed while where it is more shady an individual polyp may grow larger so that it has sufficient surface area to meet its photosynthesis needs. Food is shared and transported throughout the colony in order for the whole to grow together. This growth happens slowly: perhaps just 1 centimetre a year for the big boulder-shaped corals or up to 20 centimetres a year for the branching corals that look a little like shrubs. It can take 10,000 years for a fully-fledged reef to form from the first group of coral larvae, but given ideal conditions they can grow to an immense size. The Great Barrier Reef is the largest structure on Earth, stretching some 2,000 kilometres in length with over 40,000 square kilometres of reef.

Earth has approximately 350,000 square kilometres of coral reef, which may sound a lot but is in fact less than a tenth of 1 per cent of the ocean. This scarcity is due to the exacting requirements coral demands of its environment. Corals must have shallow sunlit waters at a perfect temperature and pH level. Any long-term changes to these conditions make it impossible for coral reefs to survive.

Corals create highly complex three-dimensional structures with a range of different-sized nooks, holes and crevices for animals of all sizes to colonise. On a healthy reef this leads to the astonishing range of species that make coral reefs so fascinating to watch, but, more importantly, it is the diversity and number of species living around the reef that keeps it healthy.

A constant danger to the corals is that seaweed might grow over them and starve them of sunlight. Fortunately corals have allies – armies of plant-eating fish like chub, rabbitfish and parrotfish that graze on the algae and keep the reef clean. Not all fish are friends of the corals, however; some, like the butterflyfish, get their nutrients by eating coral. When all is in balance, this never threatens the reef itself as there are small predatory fish like grouper also seeking a meal who are only too happy to grab a butterflyfish. However, if given a chance these groupers will also feed on the grazers. If the groupers become too numerous and eat all the plant-eaters then seaweed will smother the corals. This is why a healthy reef also has bigger predators, to keep the smaller predators in check.

It was recently discovered that on a healthy reef you can get a surprisingly large biomass of predators (both small and large) relative to prey. Indeed, there are several places on Earth where on rare occasions you get quite extraordinary numbers of sharks. This is due to spawning aggregations of fast-breeding fish coming together from multiple reefs. Fish that for most of their lives remain in the same little patch of ocean have been known to travel up to 30 kilometres for these spawning events, and the resulting bounty attracts vast numbers of sharks and other predators. Yet even in normal circumstances, a shark-infested reef is likely to be a healthy reef.

A healthy, diverse reef is, in fact, so enticing to marine creatures that it attracts new life from far away. If you put your head underwater above a coral reef you hear a sort of constant crackling noise rather like rain hitting a tin roof. It is nearly impossible for a human ear to analyse the sound but you can notice that some locations seem to be louder or more intense than others. Scientists have recorded the sound and managed to separate it into its component parts, revealing that what seems to us like radio static is actually made by a complex community all busily going about their lives while sending messages to one another – snapping shrimp communicating by firing air bubbles, fish croaking and grunting and all sorts of whistles and whoops. It is perhaps what rush hour in New York might sound like to an alien civilisation listening in from afar! So distinct are these sounds that researchers have been able to use them to distinguish between reefs in different parts of the world as well as between healthy and degraded reefs. But perhaps most importantly their research has uncovered that these sounds are used by baby coral reef fish to help them select a healthy reef where they can settle.

Many reef fish spend the first phase of their lives as plankton drifting through the ocean. As they grow, they search for a reef on which to settle. They have a set of bones in their skulls called otoliths, which move or vibrate with sound enabling them to hear, and when they hear the signature tune of a vibrant healthy reef, they are attracted to it. In this way a healthy, diverse reef entices more fish and other creatures, and in doing so stays in good condition. Conversely a damaged or overfished reef will be unattractive to nearby marine species searching for a home and thus may fall further into a downward spiral. To counter this, scientists are experimenting by playing recorded sounds to attract fish to degraded reefs, in order to help the reef recover.

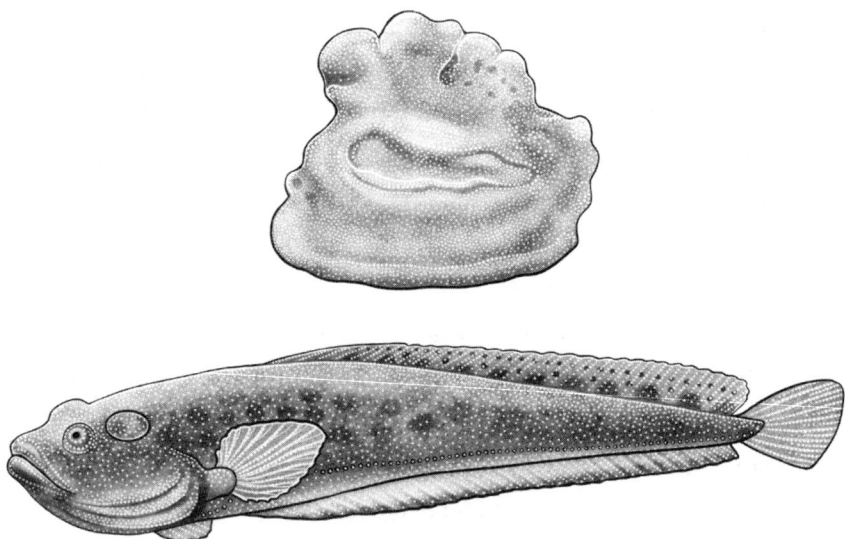

The otolith bones of an Atlantic midshipman fish. The otoliths not only vibrate with sound, but enable the fish to perceive gravity, movement and direction.

Corals, as we know them today, have lived on Earth for some 200 million years, and where they are found and in what quantity has changed with the positions of our continents, fluctuations in climate and impacts from mass-extinction events. Today you can find fossilised remains of coral in the fells of England's not particularly tropical Lake District, the Holy Cross mountains of Poland some 500 kilometres from the nearest shoreline, and even in the Spring Mountains just outside Las Vegas. However, for the last 12,000 years coral reefs have for the most part flourished between the tropics of Cancer and Capricorn, due to a change which would transform our world.

Following the Pleistocene, a geological epoch of huge climatic fluctuations and destabilisation lasting over two million years, something remarkable happened. About 12,000 years ago the Earth settled into a reliable, predictable warmer phase which has become known as the Holocene. With two polar ice caps, regular seasons in the temperate zones and warm, stable mean temperatures in the tropics, there was a balance that allowed both nature and humanity to flourish. In the dependably warm sunlit tropical waters of our planet coral grew into vast reefs and marine life abounded on them.

But now these waters are changing. We have left the safe, stable period of the Holocene – the only conditions human civilisation has ever known – and entered the unchartered and unstable period of the Anthropocene – an age defined by humans; an age in which humanity's impact on the planet is now the dominating force. The balance on which coral depends to build its wonder world is unstable, teetering and at times lurching from one state to another. Many coral reef species are so sensitive to fluctuations in their environment that a single serious impact can unleash volatile change throughout the ecosystem.

The ocean plays a vital role in moderating the globe's climate. It acts a little like a giant sponge soaking up carbon dioxide (CO_2) and heat. It is estimated that between 1800 and 1994 a staggering 118,000 billion kilograms of anthropogenic CO_2 has been absorbed by the ocean. And between 1994 and 2007 one-third of all our CO_2 emissions were taken up by the ocean. While this has saved us so far from runaway global heating, it has serious consequences for marine life. The ocean has absorbed so much of our excess CO_2 that its pH is falling – the ocean is acidifying. This change makes it more difficult for coral to grow because falling pH can dissolve exposed skeletons and makes coral more vulnerable to breakages. It also weakens the structure of the entire coral reef since much of it is comprised of limestone – the calcium carbonate in old coral skeletons – which can dissolve, thereby degrading and threatening the very foundation of the reef.

As well as this direct effect, increasing ocean acidification could cause the reef's natural attraction technique to falter by changing the sound of the reef. The shells of sea snails, crustaceans and clams are weakened by the lower pH causing their populations to decline. The reef quietens. There is less clicking of shells and snapping of shrimp. Consequently the ability of fish to hear and therefore locate a viable reef on which to settle is diminished. Juvenile fish floating in the open ocean are much more vulnerable to being eaten. Beyond the heartbreaking image of Nemo and other juvenile fish adrift in search of a home, this creates a negative feedback loop that can significantly impact the coral reefs as the next cohort of fish may simply never arrive to repopulate the reef and spawn future generations, and the reefs become even quieter

A similar end result is already occurring through a different route: fishing. Humans are creatures of habit in many ways, including the choice of fish we consume. Commonly targeted reef species include large predatory fish such as grouper and jack, which play an important role in keeping the wider ecosystem in balance. Removing too much of a single species has significant knock-on effects, particularly on an environment as dependent upon interrelationships as a coral reef. Examples of these kinds of impacts have been recorded all over the world. Coral reefs in the Caribbean with increased fishing pressure were found to have more seaweed growth because the herbivorous fish were gone. In Kenya, triggerfish were targeted by fishers and this led to a boom in population of their prey, sea urchins. Heavily fished reefs also suffer from the fishing gear knocking into corals or getting tangled in them. Polyp tissue is very thin and the underlying skeleton is fragile, so damage is easily done. Wounds are then readily infected, which weakens the colony and can even kill it completely.

While the impact of fishing is profound, there is a simple and quick solution. Countries can easily and immediately ban the most destructive and indiscriminate fishing methods, such as dynamite and poison fishing, from their waters, allowing only lower-impact methods. They can also create protected areas around the most productive or sensitive habitats. While roving pirate fishers may lose out, small-scale fishing communities will catch more fish as a result of the healthier seas and the protected reefs could attract lucrative tourism. When the Pacific islands of Palau brought in strict fishing controls initially around inshore reefs and later also on much of its offshore waters, the local fishers saw an increase in their catches as fish stocks recovered. As word got out about the incredible underwater recovery, diving and snorkelling tourism grew, bringing jobs and custom for hotels and restaurants.

Tackling coral bleaching is far less straightforward, however.

Coral bleaching happens when coral polyps are stressed by changes in conditions such as temperature, pollution, pH and salinity. As far as we know, coral bleaching has always occurred from time to time. Sudden warm periods in an area or natural fluctuations resulting from the El Niño/La Niña cycle increase or decrease water temperature for a short time around a reef or reef system. This is a threat to corals because they can only tolerate a relatively narrow temperature range. Beyond this range the photosynthetic process starts to go wrong and the zooxanthellae can produce substances that are harmful to the coral polyps. As a reaction, the polyps will expel the zooxanthellae. This is easy to see, because without the zooxanthellae the tissue of the polyp is translucent and the white skeleton underneath becomes visible – it has 'bleached'. If the temperature quickly settles back within the normal range and the zooxanthellae return to the coral then normal food service resumes. However, when the temperature stays beyond the normal range for too long, which is now far more likely due to climate change, the polyps will starve. From this there is no way back, and the coral will inevitably die.

While some bleaching events have been reported due to pollution or extreme low tides exposing the coral to intense sunlight and heat, by far the biggest cause is exposure to prolonged ocean heatwaves. They can last for weeks or even months and are even able to warm the ocean to depths of hundreds of metres. Climate change has made these heatwaves more extreme and more frequent.

The future for coral reefs is often portrayed as extremely bleak indeed, and there is good reason for such concern. The global bleaching event in 1998 that killed 8 per cent of coral worldwide has been well studied, giving us both warning and hope. In the Seychelles approximately 90 per cent of hard coral was lost. Recovery here was closely studied and researchers found that once algae had swamped the reef, it slowed the coral rebound. Where over-fishing herbivorous fish was a problem, impacted reefs struggled to recover. In many places corals managed to recover strongly and by 2010 the global estimated coral cover was back to its pre-1998 level. However, this recovery took more than a decade; with bleaching events occurring more and more frequently there is often not time for a full recovery before the next one hits. Since 2011 there has been a steady decrease in coral and an unprecedented increase in algal cover as we continue to heat our planet.

Global bleaching events have occurred in 1998, 2010, 2014–17 and at the time of writing in 2024 the world's coral reefs are undergoing another. If we continue to release greenhouse gases into our atmosphere at the current rate, then very few coral reefs will survive. But even if we succeed in rapidly phasing down our emissions there is already unavoidable warming and acidification locked into the planetary climate system. This will have serious consequences for coral reefs.

Yet, it appears that we can at least minimise that damage.

A clue to how we might do this comes from the region with the most diverse reefs in the world, the Coral Triangle. This area of the Western Pacific between Indonesia, Malaysia, the Philippines, Papua New Guinea, Timor Leste and the Solomon Islands covers 6 million square kilometres. It has an astonishing 76 per cent of the world's coral species and scientists have recorded over 2,000 species of reef fish in its waters. More than 140 million people rely on fish and seafood from the Coral Triangle either for their livelihood or as their main source of protein. Here scientists found that there had been an increase in hard coral cover over recent decades, as well as a decrease in algal cover. It is thought that the density and diversity of the coral in this area helps them to be more resistant to spikes in temperature.

Similarly, when a reef has a healthy ecosystem and fish population then it is significantly more likely to survive bleaching events. There is an excellent example of this in the middle of the Pacific Ocean. In 2009 a National Geographic Pristine Seas expedition discovered some of the most stunning reefs in existence around the remote and uninhabited Southern Line Islands. Their findings helped secure protection for the reefs through a so-called no-take zone that closed them to all fishing. Then, during the 2016 El Niño, it seemed that disaster had struck. The worst bleaching event on record hit the Southern Line Islands: over half the coral was lost. Scientists expected the reefs to suffer the same fate as many others around the world and die completely. But then something astonishing happened. The corals returned, in some places even richer and more spectacular than before. The reason? Fish. Without them, bleached coral skeletons are quickly smothered in algae and the entire reef is overwhelmed. But because of the Southern Line Islands' strict protection, armies of plant-eating fish survived and grazed the algae, allowing the coral time to recover.

Tackling climate change is a global issue, albeit one requiring national and local leadership. However, buying reefs time and helping them survive the global heating and ocean acidification that are now unavoidable can be done locally. Protecting a reef from both overfishing and pollution are within the power of local communities and local governments. And wherever that has happened the reefs are healthier, the small-scale fishing communities have more to catch and, most importantly, those benefits are more likely to continue well into the future.

CABO PULMO, BAJA CALIFORNIA, MEXICO

When you view drone footage of Cabo Pulmo recorded from a couple of hundred metres above the ground, you understand why the place is special. From that height and with a decent lens you can clearly make out the lines where four realms transition into one another. From the east the scrubby, rocky hills are an arid beige interspersed with khaki pockets of vegetation running along the valley and ravines where water periodically flows and briefly settles. A thin strip of sand reflects the sunlight and every few seconds is invaded by the white foam of breaking waves. Immediately beyond the surf lies a lagoon of vivid blue-green shallow water. A little further offshore lies the true wonder of Cabo Pulmo. Ridges running north to south separate the light, inviting lagoon from the dark, open ocean. These ridges are formed of granite, which provides the base on which coral can grow.

As the world warmed after the last ice age, ice melted and retreated, and sea levels rose and flooded continental shelves. Along equatorial latitudes this created the conditions for warm, shallow seas – excellent places for coral colonies to establish and grow. Cabo Pulmo is right on the edge of these latitudes. At the bottom of Mexico's Baja California near the mouth of the Gulf of California, it is just still warm enough for corals to grow – head much further north and warm water corals soon disappear. It is a rare and precious outpost.

While from the air you may be able to understand why this area is geographically interesting, you need to go beneath the waves to realise that this relatively small patch of a much larger community has an amazing story. It is the diversity and density of the coral which first strikes you. The rock here runs like fingers on top of the seabed into progressively deeper water, and it is this that allows for habitat variations that favour different species.

The colours, the intricacies and the sheer beauty of this underwater world lead many visitors to assume it is a pristine, untouched reef. The coral density here is among the greatest in the Gulf of California, and that helps to attract a great diversity and abundance of other marine species. The shoals of jacks here are so vast that the vortexes in which they swim have been known to engulf divers. Bull sharks, humpback whales, manta rays, turtles and sea lions all frequent the waters, and the huge array of birdlife here, attracted by so much fish, is considered to be of global importance. You can see why writer John Steinbeck said it 'pulsed with life' when he visited it in the 1940s.

But, in fact, Cabo Pulmo is not pristine. By the 1980s its pulse had stuttered and faded. A few fish, echoes of the once vibrant reef community, stubbornly remained, but the great ocean travellers – sharks, rays, turtles and whales – had not passed by in years. Cabo Pulmo had joined the long list of locations erased from the travel routes of the ocean's wandering giants.

Local fisherman Juan Castro didn't truly think he could make the sharks or whales return, but he did hope he could revive his community. The fish catch had plummeted. Decades of overfishing combined with the anchors and weighted nets of boats smashing the reefs had finally tipped the intricate ecosystem too far out of balance. Fish populations were critically low and the people were out of options.

It would be easy to give up or, as so many other places had done, use increasingly brutal fishing methods to catch the few fish that remained. But Juan Castro remembered the rich seas of his childhood, the sharks he had seen while fishing with his father and the proud community that thrived in this beautiful region. He believed in the power of the sea and he believed in his neighbours. Working with a local marine professor confirmed that what he instinctively understood from a lifetime of working the sea was also backed up by the science: if given both time and protection the reef on which his community depended could recover – it could bounce back to life.

First they studied the reef, identifying areas where some healthy coral remained and small shoals of fish could still be found. They then campaigned and petitioned the government to turn the reefs into a marine protected area. This would keep the foreign fishing boats out, but that alone wouldn't be enough. The professor's studies suggested they needed a complete no-fishing zone to allow the reef to recover. For a community largely comprised of fishing families, this was a life-changing request. Juan persuaded them it would be worth it and in 1995 the Cabo Pulmo no-fishing zone and protected area was established.

Jubilation was short-lived. Now came the tough part. Having given up their fishing rights, the community's main source of food and income had disappeared overnight. It is hard to imagine what it must have been like for a working community, built by generations of fishing families, to live on the edge of the sea yet be unable to fish and rely instead on government food vouchers to support them. The pressure and responsibility Juan and other advocates of the protected area must have felt, to deliver results, would have been intense and daily. The patience shown by the whole

A clown fish swims among the tentacles of an anemone. There are over 1,000 known species of anemone, yet only ten have evolved mutually beneficial relationships with clown fish which are vital for their survival.

A healthy coral reef in Raja Ampat, Indonesia, 2024. Corals create highly complex three-dimensional structures with a range of nooks, holes and crevices

perfect for animals of all sizes to colonise, resulting in the astonishing range of species present in the well-protected reefs of Raja Ampat.

Green star coral (*Pachyclavularia violace*) from a time-lapse shot. A single square metre of a healthy coral colony may have more than 10,000 polyps.

of Cabo Pulmo in continuing to resist fishing in those first few years until the early signs of recovery became visible is remarkable. From shore they saw the seabirds growing in numbers and diving for the fish that people were not allowed to catch. Still they waited. Still the marine biologists studied the waters.

Ten long years passed.

They could sense a quickening. A positive tipping point had been reached. Biodiversity begat biodiversity, rich seas attracted more species, healthy coral spawned and gave shelter. Sharks were spotted, whales returned.

Ten years felt like a very long time to Juan and the community of Cabo Pulmo, but for scientists the transformation was surprisingly, delightfully rapid. This scale of regeneration in such a short space of time gave hope. Few habitats on land would recover as fast. Leave a previously farmed field alone for a decade and you will see change, but nothing on this scale. The pace would, of course, vary in different parts of the world, but, for example, if you just left alone a typical lowland western European agricultural field, you might see brambles and nettles quickly take over the once disturbed land, acorns long dormant in the soil might sprout and become young oak saplings. Probably thorny scrub would dominate, created by berry eaters such as thrushes feeding on the open ground. Perhaps wind-dispersed species like ash would begin to grow if there was a woodland nearby. Over a decade you would certainly see a rise in biodiversity, particularly among birds, insects and small mammals. Carbon sequestration and soil quality would improve. But even on abandoned farmland close to existing old-growth woodland, it would be several decades before a habitat remotely resembling native woodland would return, with all the species it supports. In Cabo Pulmo, however, there was recovery across all levels of the ocean ecosystem, from the smallest reef fish to the largest predators, and that was due to the power of the ocean and the characteristics of the species that have evolved there.

It is easy to assume that a coral reef functions much like a habitat on land. Indeed, they are often called 'rainforests of the sea' due to their biodiversity. But there are crucial differences, especially when comparing the respective roles of mammals and fish. Even though many mammals and birds continue to produce offspring into old age (humans are a rare exception), they don't have more babies at any one time as they enter middle and old age – holding steady is an excellent result for most. Conversely in many fish species the older, bigger fish have significantly more eggs than when they were younger and smaller. Age-old logic required throwing back the young fish you catch and allowing them to breed but taking the older ones to eat. This makes some sense because clearly you don't want to wipe out the young pre-breeding population, but it doesn't recognise the super-reproductive powers of big fish. And after ten years of protection Cabo Pulmo had a nicely ageing cohort restocking its waters at an ever-faster rate.

Marine biologists had rarely seen anything like this before. Their studies showed that fish numbers had increased by more than 400 per cent inside Cabo Pulmo's no-take zone and the booming population had already begun spreading outside the protected area, repopulating other reefs. This is another important difference between the land and sea. In the sea there are no fences, roads or towns. So once a protected area has healthy fish populations then gradually some of the fish move beyond the boundaries of the reserve. This is known as the spillover effect and it is a crucial reason why no-take zones will, if given time, enable fishing communities to catch more fish in the surrounding waters than they did before protection was introduced.

The fish numbers provided solid proof of recovery but, for the future of Cabo Pulmo, beauty and biodiversity were also important. The burgeoning coral gardens, the curious rare invertebrates, the resident sea lion colony, the vast seabird flocks and the star power of whales, dolphins, turtles and sharks together made for a dream destination for ecotourists. This was a vital development. The Cabo Pulmo marine reserve was now so successful it was declared a national park. This excluded foreign fishing vessels permanently but allowed locals to resume setting their nets. The future of the town was entirely in local hands.

Fishing was once again a profitable business. Far more fish could be caught in less time, using less fuel than had been the case before the marine protected area was established. But the community of Cabo Pulmo did not want to rely on fishing alone. By opening dive shops, guesthouses and restaurants, together with establishing tour guide services to share the wonders of their reefs with the world, they could remain a proud fishing village without the risk of overfishing.

Juan's vision and the far-sighted determination of the community has saved both the town and the reef while providing the world with a template for a better way of fishing, conservation and managing our seas. But it may have done significantly more than that. It might even have bought us time itself. Cabo Pulmo is right on the limit of where coral can grow in the eastern Pacific. Any further north is simply too cold – for now. But as ocean heatwaves, driven by climate change, cause reefs across the world to bleach, it is the coral at the outer limits that might survive the longest. We are now all but certain to exceed 1.5 degrees Celsius of warming across the planet, but if we act fast enough it is still possible we might stabilise temperatures before they run out of control. If so, places like Cabo Pulmo could become genetic refuges of incalculable value – sites from where the coral reefs of our planet could re-emerge over time.

The miracle of Cabo Pulmo has transformed our understanding of coral reefs. It has shown that simply leaving parts of the ocean alone creates the capacity for it to regenerate, repopulate the surrounding seas and defend itself against seemingly impossible odds.

We must not get complacent. Continued global heating will eventually overwhelm all but the most fortunately located reefs. But by making wise, bold decisions about how we fish the ocean we can buy ourselves a little more time to make wise, bold decisions about how we address climate change. And, of course, in the process we can transform our coral seas and the livelihoods of those that live alongside them. This is perhaps the biggest win-win on our planet today, and we should do all we can to enable a world of Cabo Pulmos.

2

THE DEEP

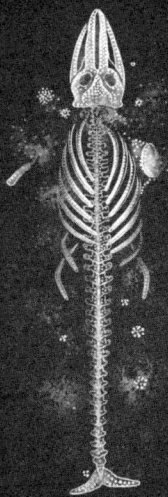

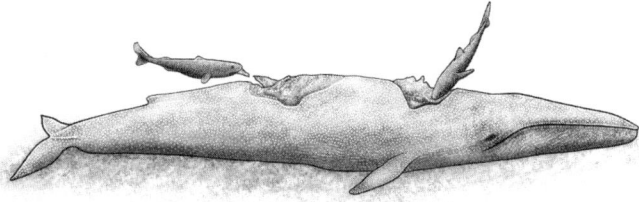

THE SUBMARINE, AUSTRALIA

Leviathan, Kraken, Poseidon . . . the deep ocean has long fired our imagination, just as much as looking up to space has always done. Perhaps there is something in the human condition that requires us to invent stories in order to make sense of the places we do not yet understand and cannot yet scientifically describe. Even the word 'abyss' comes from the Latin term meaning 'bottomless pit'. But while on a clear night almost all of us can make out the stars, and anyone with even a rudimentary telescope can inspect the landscape of the light side of the Moon, the ocean depths have for most of humanity's existence been unknowable.

As a boy I remember reading excitedly about the naturalist William Beebe being lowered into this totally unknown world inside a spherical steel ball called a bathysphere, which he had designed with an engineer named Otis Barton. Their experiments took place off the coast of the wonderfully named Nonsuch Island, near Bermuda, and they were progressively lowered further and further, ultimately reaching a record-breaking 922 metres below the surface. I used to imagine what it must have felt like to be inside a tiny structure with such immense pressure bearing down on it from the surrounding ocean. Beebe's descriptions of fleeting, half-glimpsed encounters with strange deep-sea life thrilled me, and I wondered if he would ever be able to obtain a picture of one of those creatures.

Despite the vast technological leaps in the time since, today there is still precious little film footage of the deep, and only a fortunate few ever get to see in reality the life that exists below the sunlight zone. But in my eighty-eighth year I got the chance to be one of them. In 2014 I was making a programme for the BBC about Australia's Great Barrier Reef. I had first filmed there some sixty years previously using primitive scuba gear and was now returning to use the most advanced modern technology to reveal how life on the reef functions, the threats it faces and some of the most exciting recent discoveries about the reef. One particular sequence we hoped to film would enable me to achieve all of those objectives by descending in a submarine to 300 metres, which at that time would be the deepest dive ever recorded by a craft of this kind on the Great Barrier Reef and surrounding area.

Our state-of-the-art ship, the *Alucia*, was purpose-built for scientific investigation and carried a Triton submarine, together with its expert pilot. The submarine had a high-pressure-resistant Plexiglas dome through which we would be able to film, and powerful lights to illuminate our surroundings once we descended below the reach of sunlight.

We planned to dive on Osprey Reef, which lies in the Coral Sea between Papua New Guinea and Australia, approximately 90 miles off the north-east coast of Queensland and on the very edge of a 2,000-feet drop-off. This feature would enable us to look at coral at a variety of different depths in a single dive, but our main purpose was to film and collect a sample of a mysterious community of corals at depths of between 100 and 150 metres.

In my lifetime the world has lost over half its coral. The Great Barrier Reef has been especially affected by climate change and pollution. Yet this particular deep-sea coral community was thriving, seemingly unaffected, and scientists were wondering if it could help to repopulate other parts of the reef as the impacts of climate change took hold.

The submarine that would take us there was a small, transparent, almost spherical craft. Inside there was just enough room for the pilot, a cameraman and me sitting side by side. To begin with the submarine descended quite slowly. As it sank, we had a chance to look around just above the surface – a 360-degree view against which waves were gently lapping. As we slipped below the surface, it was impossible to fight the instinct to hold your breath.

It was an awesome, slightly worrying moment.

We dropped down the face of the reef past spectacular soft corals. Shoals of pelagic fish mingled with resident species against a backdrop of gorgonian fan corals over 2 metres high. At 30 metres deep we paused to watch a green turtle. It seemed to take no notice whatsoever of our presence, even though our craft had transparent walls and we were fully illuminated as we sank past it and descended into the mesophotic (twilight) zone. We were now between worlds.

While it appeared quite dark, there was still enough light to sustain some specially adapted corals which collect such light as there is and photosynthesise in the same way as their relations closer to the surface. To do this, they have evolved wide flat plates to capture as much of the remaining light as they can. At this depth they were out of reach of cyclones and the rapidly warming shallower waters that cause coral to bleach. As a result, the scientists with whom we were working thought these corals might survive when others closer to the surface could no longer do so.

Most of the light that reaches this depth comes from the blue end of the spectrum, so the rocky reef appeared dark blue. But we turned on the submarine's lights and colourful relatives of the corals living in brighter waters above became visible. We had a mechanical arm which we used to collect a sample. Then, we went deeper still.

After 200 metres it was so black that no corals were growing.

Three hundred metres and we touched bottom. At this depth our dials showed that the pressure was more than thirty times that at the surface, an alarming 450 pounds per square inch. The lights from our submarine revealed countless particles slowly descending all around us and settling on the seabed. The marine biologist Rachel Carson poetically described this phenomenon in her 1951 book, *The Sea Around Us*, writing: 'When I think of the floor of the deep sea . . . I see always the steady unremitting downward drift of materials from above, flake upon flake, layer upon layer – the most stupendous "snowfall" the earth has ever seen.'

This has been called 'marine snow'. It is a shower of biological debris falling from higher levels like leaves from the canopy of a forest to the floor below. Much like those terrestrial leaves, marine snow is a vital part of the ecosystem and an essential base of the food chain for deep-sea creatures, which filter it out as it falls or scavenge it from the bottom. Below 1,000 metres the majority of this snow does not reach the bottom but instead is consumed by animals and bacteria on the way down and returned as dissolved carbon dioxide and nutrients in the water column. But at our present depth, some 300 metres, it was still falling all the way down to the sea floor in surprisingly large quantities. The snow that is not eaten joins the muddy sludge on the bottom that in some parts of the ocean can be hundreds of metres thick. Whether consumed on the way down or accumulating on the sea floor, marine snow is responsible for transporting enormous quantities of carbon into the deep sea each year, locking it away from the atmosphere – another of the ocean's vital contributions to our planet's stability. But it is neither uniform across the ocean nor continuous in a particular location. Its thickness is entirely dependent on what is happening above. When plants and animals, most often phytoplankton and zooplankton, die nearer the surface, their bodies drift downwards. Faecal matter, sand and soot may also form part of this downfall. On Osprey Reef, parrotfish munching coral to extract the tiny coral polyps hundreds of metres above were contributing a significant proportion of the marine snow that I could see from the submarine.

It was almost time to start our ascent when we spotted a deep-water grouper. It appeared to take a great interest in us and swam right alongside the sub for several minutes, enabling us to get a clearer look at it and record something. It was a couple of metres long and we later learnt that this species had never before been seen at these depths – although in all likelihood, given how little we know of life in the deep ocean, it was probably quite common. It was we who were the rare organisms at these depths, not the grouper.

We returned to the surface, and to light and colour.

DA

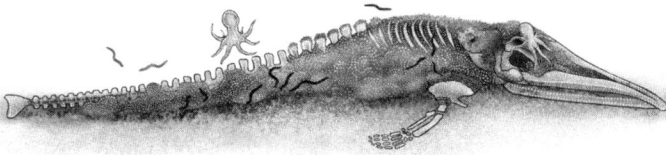

Charles Darwin's *On the Origin of Species* and Jules Verne's *Twenty Thousand Leagues Under the Sea* were published just a few years apart. Of course, *Twenty Thousand Leagues Under the Sea* was a story, never intended to represent scientific fact, but it is telling that deep sea and deep space were considered equally mysterious settings for science fiction at exactly the same time in human history that we were grasping the scientific facts of life on land. Similarly revealing is the fact that some 160 years later Darwin's work is still considered to be, for the most part, an accurate description of the evolution of life on Earth, while we are still seeking to understand life in the true deep. It is a timeline that illustrates just how hard it has been for us to persuade the deep to reveal its secrets.

In *Twenty Thousand Leagues Under the Sea* Jules Verne famously wrote, 'We may brave human laws, but we cannot resist natural ones.' Yet when he created this great work of submarine science fiction in the late 1860s, scientific study of the deep was about to challenge what we thought to be the natural laws of the ocean. At that time some thought that no life could exist below 550 metres, following Edward Forbes's 1830 study in the Azores. Based on what he had found, Forbes developed the azoic theory, which stated that without sunlight no plant could grow, and that at extreme pressure animal bodies would collapse. Logically, therefore, there should be no life in the deep. Although hotly debated by his contemporaries, this argument took place at a time when no one in the world knew what was in the deep or how it might survive there.

But a ground-breaking expedition would change all that. Following its launch in 1872, HMS *Challenger* sailed for four years, covering nearly 128,000 kilometres, undertaking thousands of experiments and discovering over 4,000 new species of marine life, including some at depths far below 550 metres. From that moment on we have never stopped exploring the deep ocean and it has never stopped rewarding our curiosity.

As more and more life in the deep was uncovered, the names scientists gave the improbable species suggest that, even then, they realised they were only revealing slivers of a still-mysterious realm: 'bone-eating zombie worms', 'vampire squid', 'bloodbelly comb jelly' and even the 'headless chicken monster' – a wonderfully strange form of sea cucumber that, unlike most of its kind which typically remain on the bottom, can use a special flap to lift itself off the sea floor and 'swim'. It is a dark red colour and looks a little like a swimming human heart or, with a bit of imagination, a plucked and decapitated chicken. Intriguingly it has an inventive escape mechanism – or, rather, it has a curious ability which we believe to be an escape mechanism since it has been witnessed too rarely for anyone to be certain. When disturbed it can leave behind a 'ghost' of itself to confuse attackers. The ghost is a layer of luminous skin that the headless chicken monster instantly sheds in order to distract pursuers and allow it time to escape.

A United Nations document issued in 2017 ahead of a global ocean conference stated that 'the ocean contains 200,000 identified species but the actual numbers may lie in the millions' – a revealingly wide window of uncertainty. The ten-year-long Census of Marine Life, an extraordinary collaboration of more than 2,700 scientists from over eighty nations between 2000 and 2010, estimated that about 230,000 species of marine animals have been described by science so far. The census itself found and described in excess of 1,200 new species, with thousands more awaiting formal study and description. At the time of writing, the World Register of Marine Species has documented 245,888 species, and on average 2,332 new species are added every year, meaning that at current rates it will take several hundred years for science to find, name and describe the estimated 1 to 2 million species that might still be down there. But the truth is that these numbers are educated guesses – no one knows how many species live in the ocean.

What we do know is that there are a lot of marine species still to discover, and that many of them are likely to be found in the deep. Some, almost certainly, will have evolutionary adaptations that will surprise and astonish us, as do the ones we have already discovered.

Fish species that evolve to live in dark caves typically lose their sight over generations, natural selection acting to divert precious resources to more useful senses. Yet in the deep ocean that is not the case. At depths no sunlight can reach, some of the species we have discovered have huge eyes relative to their body size; sight is clearly an important part of their world. In fact many creatures in the deep use bioluminescence to communicate or find prey, so require highly evolved eyes that are able to detect many more shades of blue and green (the typical colours of bioluminescent light) than humans can. While this makes perfect sense with hindsight and careful study, few would have predicted it based on our understanding of life in the rest of the ocean. And with so many of the species of the deep yet to be seen, let alone studied, we must acknowledge that we have merely splashed the surface of the evolutionary wonders contained there.

The black seadevil anglerfish. As well as their bioluminescent lure, these fish are unusual for being mostly scaleless, and instead, rather gelatinous.

In many respects it is thrilling that we don't yet know what lives in our ocean, and particularly what lives in the deep. In a world where, at the click of a button, we can see images of most of the land mass of our planet or travel in a matter of days to locations so remote that just fifty years ago they would have required a long expedition, there is something both exciting and reassuring about the fact that much of our world remains not just unknown but, with current technology and research budgets, essentially unknowable. The counterpoint, of course, is that the damage we are inflicting on our ocean is so profound that it must be extinguishing entire species or, at the very least, decimating populations before we even know they exist, and certainly before we discover their place in this marvellous, entangled web of ocean life.

It is not a failure of science that we do not truly understand the deep: it is a factor of its sheer scale combined with the challenges of marine exploration. Imagine a fisher sitting in a small boat in the middle of the ocean far from land – for the purpose of this illustration we will ignore the practicalities of how they got there. They have traditional fishing rods with a small weight attached to the end of their line. Unfortunately they have tied the weight too loosely and while struggling to reel in a fish for dinner it detaches and begins to fall. For the first few minutes it descends through the sunlit zone, the area we are most familiar with and where direct photosynthesis plays a dominant role in life. Water absorbs light very strongly, so by about 200 metres the weight will leave sunlight behind and enter the mesopelagic zone, also known as the twilight zone. While there is still blue light here – enough for some species to see by – there is insufficient to allow much photosynthesis, which is why the twilight zone is often considered the beginning of the deep ocean.

Some species have evolved extraordinary adaptations to exploit this zone. One of the most remarkable is the

The cock-eyed squid is a rare example
of non-symmetrical eyes in nature.

cock-eyed squid or *Histioteuthis heteropsis*. It has a large yellow eye, which it typically directs upwards towards the surface, and a smaller blue eye, which looks down. After many puzzling studies, scientists finally had enough recordings to reveal that the larger eye helps the squid to spot prey against the very dim light from the surface, while the smaller blue eye simultaneously looks for flashes of bioluminescence from potential prey in the deep.

The fisher's weight could fall for as much as thirty minutes through the twilight zone, depending on factors such as mass, drag and salinity. If it were falling towards the end of the day, it might pass one of Earth's great wonders. For the largest migration on our planet is not the great herds of the Serengeti or the intercontinental journeys of swallows, but instead begins in the dark depths of the open ocean, when millions of hungry siphono-phores, squid and fish – in particular lanternfish, one of the most abundant vertebrates on Earth – head towards the surface to feed. This is thought to be the biggest daily movement of biomass on our planet, although some have estimated that human commuters might exceed it! It caused great confusion during the Second World War when the recently developed technology of sonar detected the sea bottom thousands of metres closer to the surface than depth soundings proved it to be: such was their density, the sonar was mistaking the massive marine aggre-gations for the sea floor itself.

By feeding in the sunlit shallows on food chains origin-ating with phytoplankton growth and defecating or dying in deeper water, the mesopelagic species are thought to transfer significant amounts of carbon each day, locking it safely away in the water column or at the bottom of the ocean. Yet important though this daily vertical migration is, in just the last few years scientists have discovered a similar role being played by something much more surprising. Every year trillions of tiny crustaceans called copepods spend months feeding and fattening themselves near the surface before descending hundreds or even thousands of metres where they hibernate for between six and nine months. Here they live off their stores of body fat whilst the waste they emit is stored in the deep ocean, effectively transferring carbon from the surface to the deep ensuring it cannot re-enter the atmosphere and affect the climate for centuries or even millennia. These copepods are so numerous that their seasonal vertical migration is now recognised as a major route through which carbon is sequestered in the deep ocean.

The weight we dropped continues to fall. After about 1,000 metres it enters the bathypelagic or midnight zone. There is no sunlight at all here. For the next 3,000 metres, which might take the falling weight ninety minutes or more to travel through, the midnight zone endures. Across the world the upper layers of the ocean fluctuate and change temperature, but throughout the midnight zone, in any part of the ocean, anywhere on Earth, the temperature maintains a constant 4 degrees Celsius. There is no photosynthesis here due to the lack of sunlight and therefore limited phytoplankton, but zooplankton are nevertheless present – perhaps avoiding predators nearer the surface. It is also thought that the midnight zone is a critical habitat for larvae to develop before they grow to sufficient size to migrate and populate other parts of the ocean.

The average depth of the ocean is 3,500 metres, so in

many locations our fishing weight would have ended its journey in the midnight zone after perhaps a couple of hours' descent. But in other areas it would go much, much deeper: at the deepest point of the Mariana Trench in the Pacific Ocean, known as Challenger Deep after the ship whose crew first measured its depth, the weight would lie 10,920 metres below the fishing boat. In comparison, Mount Everest is 8,850 metres high and an intercontinental aeroplane flies at roughly 10,000 metres. This perhaps gives some context as to why scientists believe we have only discovered such a small proportion of the species in our ocean – especially when many are tiny or live in the sediment of the deep-ocean floor.

Over the last decade we have had much greater success in mapping the floor. Indeed, the importance of knowing the shape and structure of the ocean has been recognised in recent years and there are major efforts to map the entire ocean floor with modern equipment by 2030. There are many reasons why we need to do this, in particular to broaden our understanding of the role of vast oceanic currents in regulating global climate. We use our existing knowledge of these currents to help predict how the climate will change under different scenarios of carbon emissions, but by knowing in detail the shape of the ocean floor we can better understand how and why deep-ocean currents move, enabling us to grasp more accurately the influence of the ocean on global climate and therefore to predict what may happen in the future.

But there is much we have already discovered. The deep ocean is not, as we might have expected, an endless flat, muddy plain, though some bits are indeed flat and many areas have exceptionally deep mud. Indeed, scientists have measured mud hundreds of metres deep in parts of the ocean floor. When you think of all the carbon-rich sediment accumulated over millennia in mud so thick, it helps to visualise how the ocean plays such a critical role in mitigating climate change. Beyond this, though, the ocean floor is a highly complex environment with immense geological features, including many that we know from land – canyons, mountain ranges, undulating hills, plateaus, ridges, volcanoes – and some unexpected ones too.

In February 1977 a team of scientists discovered warm water coming out of the sea floor along a deep-ocean spreading ridge – formed by tectonic plates moving apart – in an area of the eastern Pacific Ocean between the Cocos and Nazca tectonic plates known as the Galapagos Rift. They were the first in the world to discover an active hydrothermal vent – a structure formed where cold seawater meets hot magma. The studies that followed turned deep-sea science on its head. It was previously believed that all deep-sea life – even that which never encountered sunlight – relied at least indirectly on photosynthesis. Deep-sea creatures either migrated closer to the surface to feed or relied on surface nutrients somehow getting to the bottom: dead whales falling to the ocean floor – known as whale falls; living creatures defecating in the deep; marine snow; dead kelp sinking and other such ephemeral occurrences. But in the Galapagos Rift they discovered 30-centimetre-long clams, tube worms and an entire long-established ecosystem thriving on the nutrients and chemicals seeping out of the vents.

Jack Corliss, from Oregon State University, was on board the underwater vehicle known as the *Alvin* at the time and pre-empted the astonishment of the entire scientific world when he said to his colleagues on the ship 2,000 metres above: 'Isn't the deep ocean supposed to be like a desert? . . . Well there's all these animals down here.' The crew of the *Alvin* had not only seen life on the ocean floor, they had observed an entire ecosystem that had evolved without utilising the sun's energy.

Until this point we had believed all food chains began with photosynthesis – plants, algae or bacteria using solar energy to turn carbon dioxide and water into sugar and oxygen. But the *Alvin* team had witnessed something entirely different – chemosynthesis: here food chains start with microbes creating food using the energy released from chemical reactions. Life that did not need the sun.

Since that first discovery hundreds of such vents have been located including the Lost City, an astounding hydro-thermal vent field 750 metres below the surface of the sea located a few kilometres from the Mid-Atlantic Ridge just south of the Azores. It is estimated to have been active for at least 30,000 years and has chimneys – towering solid structures formed by the minerals coming out through the vent – up to 60 metres tall. The water temperature coming from these particular vents is lower than has been discovered in other vents to date, and as such they appear to support a different community of creatures. Many scientists think these conditions may be the closest yet found to those where life originated on Earth. It is a site of incalculable scientific value.

While many of the geological features of the ocean floor are somewhat analogous to those on land – mountains, ridges and volcanoes, for example – even those habitats and resulting ecosystems are very unusual to our eyes, perhaps because the conditions that determine life in the deep are so different from the ones with which we are familiar. On land, a habitat and what lives in it is, at its most basic, determined by sunlight, rainfall, altitude and available nutrients, perhaps delivered via a river, soil or migrating species. In the deep, a dead whale falling to the bottom might create a habitat for a year or two, and vegetation and fallen trees washed out to sea and then sinking could do likewise. Beyond the established habitats of deep-sea vents or big geological features such as seamounts, life is more ephemeral than it is across much of the land or coastal seas. But it is also much slower growing and longer lived.

In the deep, life has a different rhythm. There is less oxygen and far fewer nutrients or reliable food sources than in the sunlit shallows, so everything tends to go slowly – including cellular processes. This is why many species here grow gradually and live for a very long time. Interestingly, the method for determining the age of fish is similar to that of trees: scientists count the rings, or annuli, within the hard structures of a fish, such as the otoliths in the ear. A new ring is laid down each year, so by counting them under a microscope it is possible to get a good idea of age. The rougheye rockfish (*Sebastes aleutianus*) lives for over 200 years at depths of more than 800 metres off the coast of Alaska; a deep-sea coral colony (*Leiopathes glaberrima*) brought up from over 400 metres under the ocean near Hawai'i was estimated to be more than 4,000 years old.

Perhaps most infamous of all is a long-lived deep-ocean species typically found in the Atlantic and Indo-Pacific called the orange roughy (*Hoplostethus atlanticus*), or slimehead, as it used to be known. The unflattering name comes from a remarkable adaptation which allows the orange roughy to sense its surroundings in the deep. The slime comes from a mucus-producing canal on its head and is an excellent vector for conducting vibrations in the water, helping it to hunt prey or avoid predators. Orange roughy can live for over 150 years and they don't breed until they are between twenty and forty. In comparison, sardines reproduce when they are one or two and rarely live for more than a decade. These are two very different types of fish, so in some respects the comparison is unfair, but what connects them is the fact that both have been fished industrially. This makes lifespan and reproductive age highly relevant, because while life grows slowly in the deep, fishing has grown fast.

Orange roughy can grow to up to 75 centimetres in length but are more commonly around half a metre long, and they aggregate in huge shoals between 180 and 1,800 metres beneath the surface. Their social nature has been their downfall. In the late 1970s fishing technology was nothing like as advanced as it is today, but even so it had moved a long way from pole and line. Sonar, early computer navigation systems and industrial-scale trawlers with large freezers, hydraulic winches and nets miles long meant that the factory fleets could find and haul in the orange roughy in astonishing quantities. Such was the bounty and so fast did the industry grow that in just a few years the roughy went from being almost completely unknown to a staple fish in household freezers and restaurants. At first it was hard to measure the damage: ocean science takes time while trawling for highly lucrative species was taking place with a gold-rush mentality. The scientists weren't helped by the roughies' tendency to shoal: even as total numbers fell, catch sizes remained high because the remaining shoals were still easy to locate. But you can't take thousands of tonnes of a slow-growing fish out of the sea in just a few years without consequences. Many fisheries collapsed and the orange roughy has become the most famous cautionary tale of the deep.

In the process of catching epic quantities of roughy in immense trawl nets, countless other deep-sea creatures were also killed, and the seamounts around which the roughy aggregate had their fragile ecosystems destroyed. Sponge beds and delicate, deep-sea corals that had taken thousands of years to form were gone in an instant. Their rate of growth is so slow that recovery in any timescale meaningful to humans is impossible; in all practical terms they have already been lost for ever.

The plight of the orange roughy may well have mobilised scientists and conservationists to both understand and better protect the deep, but it has not dimmed the desire of some industrial fleets to exploit it. Deep-sea fishing occurs across much of the high seas. Seamounts are trawled in every part of the ocean and marketing teams rush to rebrand unappetising-sounding deep-sea fish as advertising-friendly products: the Patagonian toothfish becomes the Chilean sea bass and deep-sea gulper sharks are not mentioned at all when their liver oil is listed as squalene in beauty products.

Perhaps the strangest aspect of deep-sea trawling is how little money it makes and how few fish it yields – the orange roughy was an exception with vast fortunes made, albeit for a very short period of time. But for the incalculable damage that deep-sea fisheries inflict on our ocean, they catch less than 1 per cent of the fish we eat. Many of the deep-sea fleets are subsidised by governments or use modern slavery to keep labour costs low enough to turn a profit.

It might well be the case that a few deep-sea fisheries catch modest amounts of fish sustainably, but you cannot protect or manage an ocean based on a small number of responsible actors: you must look at the impact of the industry as a whole and act accordingly. This is especially so when most of the deep-sea fishing takes place in waters beyond national jurisdiction and therefore not managed or controlled by any country.

In our lifetimes we haven't just discovered life in the deep; we have fundamentally changed our understanding of how life can exist on our planet. The idea that life on Earth began in the deep is no longer outlandish science fiction but worthy of serious study. Although we still know very little about this realm, already we realise it is of vital importance to life across our world. At such an early stage in our discovery and understanding, surely our first responsibility is simply to leave it alone until we know more.

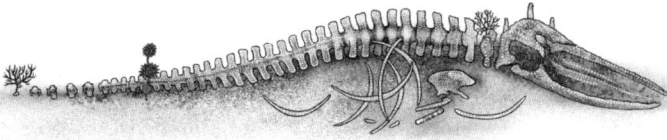

RACE FOR THE DEEP

It was as if the starting gun had been fired for a most unusual race.

A marine scientist was directing a remotely operated vehicle (ROV) the size of a small lorry 3,000 metres below his expedition ship. Using a joystick not unlike that used by video gamers, the pilot steers the ROV while a colleague sitting next to him uses their controls to move and focus the on-board cameras. They are searching for an octopus garden where over 1,000 of these usually solitary creatures have been known to gather.

In the South China Sea, far from land, another research team was carefully manoeuvring three dead cows. They lower one cow to a depth of 650 metres, a second to 1,600 metres and the third to 3,400 metres. They are on the edge of a seamount and need to get these depths exactly right – simply dropping the bovines off the side of the boat isn't going to work. A human-occupied vehicle (HOV) named *Shenhaiyongshi* is required to position the cows.

Alongside, they set up steel-framed stands each with a stills camera, a video camera and several other oceanographic monitoring instruments. They leave the cows and sail back to shore.

Meanwhile, a Canadian mining company had formed an unlikely partnership with a small island state of just 12,000 citizens, with no capital city and only one road. The mining company wants to extract tens of thousands of potato-sized nodules potentially containing rare and increasingly valuable minerals from the deep ocean floor. The island state needs money. Together they're attempting to exploit an obscure legal loophole to fast-track a new extractive industry into existence while an unlikely coalition of scientists, campaigners, governments, fishing companies and indigenous leaders race to close it down.

Suddenly, deep-sea exploration appeared to have acquired a new urgency. Less than forty-five years had passed since the famous 1977 discovery of deep-sea vents, but already humanity was in a race between those seeking to understand and explore this alien environment and those wishing to mine its newly discovered minerals. By this point we had only seen a tiny fragment of the deep ocean floor and already it was not just challenging our understanding, it was confounding our imaginations. And the more we looked the more we realised our own limitations.

It could be likened to presenting a physicist from the late 1950s with a modern smartphone. They would be able to understand aspects of it such as a microphone, a speaker or the fact the device was being powered by electricity. They might be aware of the silicon chip and be able to picture it smaller and more powerful. They may even make the imaginative leap to realise that the then embryonic technology of satellites could in the future enable us to make calls with a portable device. But they would lack the decades of connections required to imagine the AI-driven functions that billions of people routinely access as soon as they turn their phone on and connect to the internet. Similarly, in the early twenty-first century we were deploying only a rudimentary understanding to piece together our fragments of insight, uncovered in small patches of the deep ocean, in a bigger picture of life in the depths. Frequently the new discoveries were so surprising that what they actually revealed was that we lacked the basic framework even to imagine what might fill the blanks in our knowledge. A particularly remarkable example of this occurred when a US National Oceanographic and Atmospheric Administration scientist took an ROV on a routine dive to explore a seamount.

Chad King clearly loves his place of work. Monterey Bay National Marine Sanctuary in California is his passion, and when he talks to fellow scientists, schoolchildren or the TV cameras he needs little encouragement to describe the wonders that live there. Much like the field of ocean exploration, Chad started near to shore and gradually headed deeper. He began by studying life in the rich kelp forests along the coasts of Monterey Bay, but when offered the opportunity to lead an expedition exploring the deep ocean floor he knew immediately where he wanted to go – Davidson Seamount.

Davidson is an extinct underwater volcano that last

David Attenborough prepares for a 300-metre submarine dive at Osprey Reef in the Coral Sea between Papua New Guinea and Australia.

This beautiful siphonophore was photographed by a National Oceanic and Atmospheric Administration (NOAA) expedition in 2016 close to the ocean floor on an unnamed seamount just outside Papahānaumokuākea Marine National Monument, Hawai'i.

A whale fall, discovered by NOAA's Monterey Bay National Marine Sanctuary and Ocean Exploration Trust at a depth of 3,000 metres near Davidson Seamount in October 2019. Experts think that the whale had probably sunk approximately four months prior to this image being captured.

An estimated 20,000 brooding female pearl octopuses were discovered near Davidson Seamount in October 2018. Scientists believe they were found in large clusters there because of the warm thermal vents in which the females laid their eggs. Warm water speeds up the development of the eggs and they may hatch up to five times faster than those laid in ambient water.

erupted approximately 10 million years ago. It was added to Monterey Bay National Marine Sanctuary in 2008 after scientists from Monterey Bay Aquarium Research Institute discovered sponges the size of sofas and 3-metre-wide corals on the summit of the seamount – which although it rises 2 kilometres from the sea floor is still more than 1,200 metres beneath the surface. Chad knew there had been surveys of the summit and flanks of the seamount, but little was known about its base.

Chad uses an analogy to convey the uncertainty of this type of deep-water exploration. He says:

Imagine you were dangled on a rope from a helicopter on the Island of Manhattan, New York. You knew nothing about the island and it was in perpetual darkness. You had two bright lights you could wave around but that was all. You knew you were somewhere on the island but just imagine giving a description of Manhattan based only on what you saw in your lights. One time you might be dropped in the middle of Central Park, another time China Town, another near Wall Street. You can't characterise the entire island of Manhattan from the glimpses you get.

And that's what it's like on a dive – each time there is something new. Each time you get a bit more of the picture. And by the way Davidson is much much larger than Manhattan! – twenty-six miles long by eight miles wide.

No one had seen the base of Davidson before – this was true exploration. Chad's colleagues warned him it was likely he would only find some sedimented habitat with a few small corals and sponges. It would be a little bit boring, he was told. In fairness, he agreed. That was indeed the most likely outcome, but the whole point of exploration was to look where no one had looked before and the deep had been surprising explorers for quite some time.

It took three hours for the ROV to reach the ocean floor. After thirty hours roving around the bottom and finding pretty much what he expected, Chad entered the last hour of the dive before he had to bring the ROV back to the surface. The ROV was travelling across a hill when Chad and his colleagues on the surface caught sight of something through the cameras about 150 metres to their left. It was only a glimpse but it just looked a little strange – a bit like a cluster of footballs. They navigated over to it and to their amazement found twenty to twenty-five octopuses all in an inverted position. Aspiring marine biologists had been taught that octopuses were typically solitary creatures, rarely coming together, but here they were in a big group, all upside down and kilometres below the ocean surface. It wasn't just a rare or unlikely sight; it was something no one had ever before even imagined might exist. We simply did not have enough of an understanding of life in the deep to adequately imagine what it was possible to discover.

The crew speculated why the octopuses might be there. No one had a strong idea until they noticed a slight shimmer in the ROV's lights in the water column – a little like the haze you get coming off the pavement on a hot day. There was warm water coming out of a crack in the sea floor. Chad had heard of this phenomenon but there was nothing to suggest it occurred in this region – the volcano was long extinct. But he had to accept what he saw. Then, as the ROV's cameras zoomed in, they saw egg capsules underneath each octopus, and it dawned on them that these octopuses must be female.

Chad decided to take the ROV further down the slope to see if more pressure and warm water was being released there. The lights inched forwards and gradually revealed hundreds and hundreds of octopuses, all in the same inverted position, all female, all with eggs. Chad realised they were either lined up along these cracks or gathered in depressions on the floor where there appeared to be pools of diffused warm water. He was astonished by their number, but, in his Manhattan analogy, he had no idea whether this phenomenon only occurred in this small block or if it was normal for miles around.

The answer would have to wait; the ROV needed to return to the surface. Reviewing the footage afterwards, Chad estimated there were well in excess of 1,000 brooding octopuses in this small area. But technical problems meant it would be months before he could return.

This time, however, he was able to swap his ROV for a submarine. He wouldn't have to watch from the surface – he could see for himself. But what would he find? Were the octopuses always there? Was the assembly ephemeral? Was it seasonal? Chad was aware of the metaphorical as well as the literal pressure – they had mounted a huge research expedition to a remote underwater volcano and there was no way of knowing if their subjects would even be there to study.

Almost six months after the initial discovery, Chad descended over 3 kilometres to the base of Davidson Seamount. Everyone was nervous. GPS doesn't work properly in the deep sea, so while they could take the ship to the exact location on the surface, the uncertainty range grows the deeper you go and by the bottom the uncertainty range had reached 70 metres. This may not sound like a lot, but when it is pitch black and your lights only illuminate a few metres in front of you, it is far from straightforward. Remembering the exact look of a rock or angle of a slope is a vital skill for deep-sea exploration.

But the octopuses were there. In fact, over the course of the expedition they found thousands more of them all across the area. As before, all were female and all had eggs. This was an immense assembly with an incalculable number of eggs. They even saw one hatch, Chad filming as the baby octopus burst from the egg sac, rose up from the ocean floor and drifted away on the ocean currents.

They were beginning to understand this aggregation. One of the evolutionary wonders of octopuses is that females typically lay their eggs, stay with them without eating or moving away and often die soon after the eggs hatch. It appeared that by brooding in this warm deep water the amount of time from egg laying to hatching was significantly reduced. Indeed current estimates are that this warm water reduces brood time by 90 per cent.

The site was named Octopus Garden, with a nod to the Beatles song, as the scientists thought the upside-down octopuses resembled flowers and were, in effect, also growing octopuses.

As with most deep-sea discoveries, the answers prompted more questions. Where do the babies go? Too light to fight the deep-water currents, they must be swept away. But how far? Do they return to this spot as adults to brood or do they just go to whatever deep-water warm seep they are nearest to? How do they locate the seeps? Do octopus mothers down here with their greatly reduced brooding time no longer die soon after their eggs hatch, as most of their kind do?

The most pertinent question for Chad was whether Octopus Garden was a one-off global natural wonder that he had happened to stumble across, or whether there are sites like this all over the world. With the evidence they had already gathered, they might be able to protect Davidson from mining or deep-sea fishing, but what about everywhere else where these sites might occur? After many more months of desk-bound geological and hydrological study, they had identified areas where similar environmental conditions were likely to occur. But would there be octopuses?

The geological feature they had mapped at Davidson is known as a low-temperature hydrothermal ridge flank system. Its base is porous basalt volcanic rock. Cold seawater percolates through the rock and deep down into it. It then travels along underground channels in the rock. For much of its journey the volcanic rock is buried under hundreds of metres of sediment, so the water cannot escape. During its time deep underground the water warms. This is not due to volcanic activity – Davidson is long extinct – but rather because deeper into the Earth, closer to its core, the ambient temperature is warmer and more consistent than that of the circulating deep water – it is the same reason why ground-source heat pumps work to warm homes and office buildings. The water heats to about 10 or 10.5 degrees Celsius – which may not sound particularly warm to us but is much warmer than the rest of the deep ocean, which is typically around 1.6 to 1.8 degrees Celsius. The water continues to travel along these channels until it comes to other areas of lower pressure where it can rise to the surface again.

They mounted their third expedition. This time they went back to using an ROV, because while Chad and the team would have to watch from the surface, the ROV could stay down far longer than a manned submarine. After an initial study of the Davidson area they wanted to look further afield. Chad knew of a large volcanic cone about 10 kilometres away. He speculated that since it was essentially porous volcanic rock rising from the sea floor without much sediment cover, perhaps warm water would be rising to the surface and perhaps therefore it might be another octopus garden – or 'octocone', as the team had dubbed it. They debated taking the ROV back to the surface as it would be faster to move it mounted on the ship, but in the end chose to manoeuvre it along the bottom while still tethered to the vessel above. This turned out to be a very good decision indeed.

While crossing the muddy sea floor from the Octopus Garden to the volcanic cone they encountered something hardly anyone had ever seen before. A whale fall. We usually only hear about the rare occasion when a dead whale washes up on a beach but it is far more likely that a dead whale will sink to the depths. Finding one is quite another matter, however. Since more than two-thirds of the Earth's surface is covered by ocean, locating a dead whale in its depths, especially when deep-sea exploration is so difficult and expensive, is nearly impossible. Yet such falls are thought to be among the most important events in the deep ocean.

A dead whale is a bounty of nutrients for creatures of the deep. On the rare occasions scientists have come across a whale fall, it is always covered in diners. Often those scientists discover species totally new to science (but presumably common to the deep) on the carcass, drawn to a bounty of food that has suddenly arrived.

Imagine Chad's surprise and delight when the ROV's headlights made out the jaws and baleen plates of a large dead whale. A few seconds later and the whole body was illuminated. Large fish called eelpouts were stripping the remaining blubber from the skeleton, twisting and ripping it off; cod-like deep-sea fish called grenadier wriggled in and out of the ribs; crabs foraged inside the carcass; bone-eating *Osedax* worms were starting to digest the lipids from the bones; and, of course, there were deep-sea octopuses – perhaps twenty – on the whale, another dozen close by.

The first whale fall was discovered in 1987 and since then less than thirty have been found and described in scientific journals. Most of these had sunk a long time before and were just bones – albeit bones being eaten by creatures never before recorded by science. Discovering a whale fall is very rare; discovering one still with flesh on it essentially by chance was unimaginable good fortune.

Chad's ROV dive was live-streamed. The footage itself is fascinating, but it is the palpable excitement of the scientists as they go from fairly sober academic commentary to 'Oh my god, it's whale fall!' that brings home just how unusual and unexpected this encounter was.

Their best estimate was that the whale carcass had been down there for around three to four months. It was in the mid-stage of ecological succession for a whale fall. From the limited data scientists have collected so far, they believe there are four successional stages of fauna that feast on the whale. First to arrive are scavengers, such as sharks, hagfish or octopuses, that are free swimming and able to cover significant distances. They consume the soft tissue. In time opportunists such as crustaceans and polychaete worms colonise the dwindling carcass. The third stage, known as the chemoautotrophic stage, is when most of the soft tissue has been eaten and *Osedax* worms and other organisms break down the lipids from the bones. The whale the team had found was between the second and third stage. The final stage hasn't been witnessed, but scientists believe there is a 'reef stage' where the mineralised skeleton is colonised by filter feeders. These four stages overlap with each other and vary based on the size of the whale and the depth to which it falls.

In truth, little is known, though scientists are confident that whale falls are vital deep-sea ecosystems. They speculate that the precipitous decline in whale numbers during and after commercial whaling must also have been devastating for deep-sea life. But there is still much to reveal and understand.

Since most of the whale falls so far have been sighted in the east Pacific Ocean along the California coast – in all likelihood simply because this is where most of the exploration has been to date – when, in March 2020, a team of scientists from China's Southern Marine Science and Engineering Guangdong Laboratory discovered a whale fall in the South China Sea there was great excitement, both within the scientific community and among the general public. The scientists placed a camera at the site which took photos every day for two months, and the results raised fascinating questions about life in the deep ocean. Would the creatures recorded on a whale fall in the South China Sea be markedly different from those on whale falls in other parts of the world? Were any of these creatures distinct species from those found elsewhere in the South China Sea and therefore potentially dependent on whale falls? How do the mobile species find whale falls and how do they survive in the periods between whale falls? At a point in time when we know so little about life in the deep yet are already fishing it heavily and potentially about to industrially mine the sea floor, answers to questions like these are vital to discover.

Which brings us back to the cows.

The researchers had three options: try to discover more whale falls; wait for a dead whale to wash up on a beach and transport it to a deep-sea test site; or, devise a suitable alternative experiment. While they are quick to acknowledge a cow is biologically quite different to a whale, they felt there were enough similarities to run a fair experiment. Interestingly, our intensive livestock farming and transportation of meat around the world mean it is not uncommon for dead live-stock to end up in the sea, so perhaps the cow experiment isn't as exceptional as it appears. Indeed, in 2019 nearly 15,000 sheep drowned when the boat transporting them overturned in the Black Sea. Quite what happened on the sea floor as a result is anyone's guess, but a spike in the population of bone-eating worms is a likely consequence of this grisly incident.

They placed each of the three cows at different depths on the slopes of the Zhongnan Seamount, partly to moni-tor whether depth variation led to changes in the speed of ecological succession but also to observe whether they would attract different communities. They installed cameras on frames nearby and waited. After a few months there were already some clear trends. The shallowest cow, at 650 metres, had half its soft tissue removed in just twenty-four hours and in a few months had gone completely. In contrast, the deepest cow, 3,400 metres down, still had a third of its soft tissue remaining. The analysis of the different species visiting the three cows suggested very different communities at each of the three depths and recorded many species never before seen in the South China Sea.

More than 95 per cent of Earth's biosphere is deep ocean, and on each occasion we investigate, the deep reveals a little more. In just the last few decades we have discovered that the world's largest migration rises from the deep ocean to the surface each night; we have found hydrothermal vents which change our understanding of biology, and revealed that the deep is fundamental to stalling climate change by sequestering vast quantities of carbon and absorbing heat in the water column. The more we look, the more important we realise the deep is to life on Earth.

The low-temperature hydrothermal ridge flank systems where Chad found the Octopus Garden are now thought to be worldwide. If correct, there would be a huge area of potential octopus gardens or indeed warm water refuges for other phenomena we have yet to encounter or even imagine.

Much of what we have discovered in the deep has scientific value and some has global value for our planet's stability, but one controversial discovery has economic value. We have found rare earth elements in the deep. In particular demand are polymetallic nodules; occurring in huge quantities across the ocean's flat abyssal plain, these roughly potato-sized lumps contain manganese, nickel, cobalt and copper. Some mining companies are pushing hard to gain permission to mine these deposits on an industrial scale, and scientists are deeply concerned about the impact such activity could have on deep-sea life, carbon storage and fish stocks. We know so little about how the deep connects to the rest of our world that it is currently impossible to predict all the impacts.

The race to mine the deep raises an important question: if you were to protect areas of the deep ocean then how would you decide where to do so? Until the year 2000 no one knew about the Lost City Hydrothermal Field in the Atlantic Ocean, an extensive range of vast towering chimneys some 60 metres tall harbouring countless species new to science, and there are thought to be many similar sites yet to be discovered. Octopus Garden wasn't discovered until 2018. We still barely understand the importance of whale falls on life in the ocean; indeed, ironically, it is likely that the greatest number of whale fall sites are along shipping lanes since many whales are killed from ship strikes. And how do you protect the immense migration of lanternfish from the ocean depths to the surface unless you treat the high seas and deep ocean as a sanctuary?

Far-sighted people came together decades ago to recognise that Antarctica, the only continent on Earth with no indigenous human population, should be safeguarded from exploitation or ownership – a common place for science and peace in an increasingly competitive world. Perhaps we now have enough evidence to suggest that it would be beneficial to all humanity if the same foresight were applied to the deep.

Jessica Battle, Kaja Lønne Fjærtoft and Sol Kahoʻohalahala ('Uncle Sol') all certainly think so, and all three realised they had little time to act in the race to stop deep-sea mining. It is a race that has brought together indigenous leaders, policy experts, scientists and environmentalists from around the world. But given the odds and the time frame, you might expect them to cut weary figures, tired of fighting better-resourced opponents against the odds. In fact the opposite is true; these advocates from very different worlds are galvanised by the fact that this particular fight might now be coming at just the right moment.

Uncle Sol cut an unlikely figure as he stood to speak at a meeting of the 2023 International Seabed Authority. The suited diplomats looked up from their laptops and reams of paper to listen to a white-haired man in his early seventies wearing a traditional Hawaiian shirt speak with great passion about the Pacific Ocean. He looked like a man who had spent his life by the water and he spoke like a leader who carried generations of wisdom within him. As a former State Representative, Elder of the Papahānaumokuākea Marine National Monument Reserve Advisory Council, and Cultural Community Member of the Pacific Remote Islands Marine National Monument, Sol Kahoʻohalahala has the expertise, experience and authority to argue against deep-sea mining on any number of grounds. But he didn't start with statistics. He did not speak of 'revenue' or 'catch' or 'sequestered carbon', though he well understands the value of all these things. Instead he sought to bring greater context to what was in danger of becoming a purely geopolitical debate. He spoke of the Kumulipo – the old genealogy story of Hawaiʻi; of how the first life form to be born is from the deepest source of the darkest seas. The first coral polyp emerges followed by all the other creatures. Imagining the total destruction of the revered seas of his people, he said, 'they are decimating sacred and cultural connections for profit'.

Jessica Battle, an ocean policy and governance expert who has spent a lifetime negotiating treaties, speaks of how fifteen years of her life have been spent on the recently signed United Nations High Seas Treaty. But she knows that without big multilateral legal agreements, campaigners will have to fight every short-sighted attempt to exploit the ocean, and soon every application for deep-sea mining, on a case-by-case basis. She feels, however, that the tide is turning.

'Even the trawling industry oppose this! People I've spent my life arguing with on fisheries think deep-sea mining is too damaging for marine life!' she says. The business argument is an important one as proponents of deep-sea mining argue we need these minerals for batteries, computer chips and even wind turbines. Yet many companies that produce those products have already announced they won't purchase minerals obtained from the seabed.

She has had a front-row seat at many global negotiations on ocean management, and on this occasion is optimistic. Thirty-two countries, at the time of writing, have already paused, banned or called for a moratorium on deep-sea mining and there are serious calls for a full global moratorium. Jessica points out that, unlike many damaging extractive industries around the world, this industry hasn't yet properly started. If we put the brakes on it now, at least while we learn more about the deep sea, there are no livelihoods at stake, no businesses to close down, no green transition to manage.

Beyond international treaties, the debate is also raging in several coastal nations. Kaja Lønne Fjærtoft, a former Norwegian government employee, now works for WWF leading a coalition in opposition to her government's recent announcement that it intends to allow mining not just within its inshore waters but also as far out as the continental shelf. There are many reasons to take the precautionary principle with respect to deep-sea mining, but Kaja's coalition of universities, youth groups, scientists and NGOs has highlighted the most compelling one in their argument to the Norwegian government. Since we know so little about the deep sea, how can we possibly predict – much less manage – the impact of so much noise, excavation and pollution? In fact one of the few things we can say for certain about the deep sea is that it takes much longer to regenerate than other parts of the ocean.

The deep functions on a completely different timescale to the modern human world: growth takes centuries; habitats form over millennia; restoration takes so long it is essentially unknowable. Perhaps, then, our approach to the deep requires not just a change in policy but also a change in perspective – one that recognises our own limitations and natural tendency to focus on the short term over the intergenerational. For the deep, as with many other aspects of the natural world, there is much to be learnt from indigenous peoples.

For Uncle Sol and his people, the ocean is their country. For thousands of years they have seen no division between sea and land, and the deep has been their sacred place of creation. Thus deep-sea mining in the sacred waters off Hawai'i is not just exploitation, it is desecration. The value to them cannot be recorded in balance sheets or offset against restoration somewhere else; it resides in the connection between humanity and the ocean, in a knowledge that we can only thrive when the ocean thrives.

The deep ocean is our last true unexplored wilderness, the remaining place where we can discover things about our world that are beyond our capacity to imagine; a common area owned by no one yet vital for all life. Perhaps for the first time this is a chance to demonstrate what we have learned as a species. We can respect the rights of indigenous peoples, we can be led by scientific evidence, we can cooperate for the good of all and define value as something quite different from money.

3

OPEN OCEAN

FILMING THE BLUE WHALE,
GULF OF CALIFORNIA

B lue whales are perfectly adapted for ocean voyaging. Their powerful yet streamlined bodies enable them to travel unseen for thousands of miles each year through the open ocean. But in certain places, and at certain times, they come quite close to the shore in order to give birth and to suckle their young. One such place and time is the Gulf of California during the winter months.

It was there, in 2001, that we went to try and film a blue whale for a series called *The Life of Mammals*. The first programme was to tell the story of mammals' colonisation of the seas, and the blue whale was obviously the most spectacular example we could use to illustrate how they have done so.

Even today, no one would describe the blue whale as easy to film. But almost twenty-five years ago, it was far more challenging. There were no drones to launch within seconds from a boat; nor were there satellite tags to alert you to a tracked whale's location. We had to rely on keen-eyed spotters on the shore and hope that a light plane guided by them could fly to the right place in time to capture an aerial view of a whale swimming alongside our boat, to give viewers an idea of its huge size.

At that time, scientific data about the species was much more limited than it is now. What has become a long-term study of the Gulf of California's visiting whale population was then in its infancy. The decades of painstaking research into their migratory routes still had many gaps. And, of course, the beginning of the new millennium was only fifteen years after the ban on commercial whaling had been agreed. So the total number of blue whales in the ocean was approximately 5,000 – only 2 per cent of their natural level.

To add to the complexity of our filming ambitions, a blue whale has a prodigious ability to carry oxygen. This enables it to remain below the surface for up to an hour and then reappear so far from its original point of descent that predicting where it will be is all but impossible. Its ability to do this is thanks to the fact that its lungs have over 800 times the capacity of our own, and can expel 90 per cent of the spent air and then, in the same breath, replace it with fresh. Land animals, including humans, can only manage to do that for around 15 per cent of their lung capacity. The blue whale also has the ability to store oxygen not only in its 10 tonnes of blood but also in its tissues; this is particularly valuable when diving to depths at which the pressure is so great that a whale's lungs temporarily collapse. But that ability does make life rather difficult for anyone who is attempting to get the perfect shot of it breaking the surface.

Fortunately for us, a blue whale often dives in a repeated

cycle – first, five short shallow dives, followed by a longer deeper one. Mind you, their cruising speed when at the surface is 20 knots, so you may still lose sight of one of them during its so-called shallow dives.

To these difficulties we added a challenge of our own. The shot we really wanted was one in which, as I was speaking in a small inflatable launch, a whale would break the surface alongside me so that both it and I appeared in the same frame, and thus give as vivid an idea as possible of just how gigantic an animal it is. But recording such a shot would not be easy. It would require an expert crew, patience – and luck.

Early one morning, we left harbour and headed for the open bay. Our pilot guide in his slow-flying aircraft appeared overhead and started to circle several hundred feet above the ocean. He had explained to us how he could distinguish the spout of a blue whale from that of other species – it shoots up to 10 metres in a relatively straight jet, and the height, volume and sheer power make it hang in the air for far longer than a spout made by any other kind of whale. Once he had spotted one, he would tell us on the radio which way to go in the hope that we could catch up with it before it dived again.

After several attempts we managed to do that. As soon as we were within 20 yards of it, we pushed a small inflatable launch over the side. I jumped in, tied myself on, and within seconds we were above the whale as it cruised 20 feet or so below the surface. The blue water was crystal clear and I could see its dappled blue flanks and its great fins wafting slowly back and forth. The cameraman on the main boat only a few yards away gave me the thumbs up. I looked into the depths at the animal's back as it neared the surface. 'It's a blue whale,' I shouted excitedly over the noise of our outboard engine, and a great spout of water shot into the air and fell, drenching me. It was one of the most thrilling moments of my life.

There are many facts that one can quote in an attempt to convey the sheer evolutionary wonder of a blue whale. Its major blood vessels are wide enough for a human being to swim through; its tongue weighs over 2 tonnes; during the breeding season it can eat 500 million krill a day; and its 200-kilogram heart is so powerful that a single beat can be detected by another whale over 4 kilometres away. But the most compelling fact, for me, is one that needs the least explanation. Quite simply, the blue whale is the largest animal on Earth, far bigger than nearly all of the dinosaurs. Such a massive animal could only live in the sea, for no bone could ever be strong enough to support a body of such a size on land.

Seeing an animal so powerful and so perfectly suited to its environment would delight anyone with a love of the natural world. But cruising beside it had a deeper resonance for me. For most of my life whales had been hunted, principally for their oil. Their numbers had fallen to such a low level that I, like many others, feared that they would never be able to recover. Even after the ban on commercial whaling in 1986, I worried that it would be a very long time indeed before we would see much of a population rebound. After all, as far as we can tell, blue whales may not reach sexual maturity until they are ten to fifteen years

old, gestation takes nearly twelve months, and they probably only calve at best every two to three years. Putting
such biological facts together with a recognition of the
vastness of the open ocean and their wide geographic distribution made even seeing one blue whale so soon after the
cessation of hunting truly remarkable and unforgettable.

But witnessing the recovery of the blue whale over the
last forty years from just a few hundred in 1982 to between
5,000 and 15,000 today testifies to the importance of
global cooperation, the power of the ocean to restore its
numbers – and the indomitable spirit of those who fought
so hard and with such success to protect these majestic
creatures.

DA

A leatherback sea turtle hauls her way out of the warm
Pacific water and on to a beach in Costa Rica. The oldest
and largest of all turtle species, her 2.4-metre-long body
has evolved to be graceful in the water, her long, powerful
front flippers enabling her to swim immense distances and
dive hundreds of metres. She now uses these same flippers
to drag herself across the beach, although, without the seawater's buoyancy, moving 500 kilograms of turtle across

sand is hard work. But it is a vital task. She is the latest in an ancient line to visit these beaches to lay her eggs. Once laying is complete, she must feed to regain her strength. She then travels thousands of kilometres down the coast of South America, heading out via the rich waters of the Galapagos, searching for the upwellings where she will feast on jellyfish. She is a member of the Eastern Pacific leatherback population and every two to four years all females of this group will make a similar journey. Others of her kind will nest in Indonesia and cross the Western Pacific to feed off the coast of Australia or in some cases even continue all the way to the west coast of the USA, annual journeys of over 13,000 kilometres between their nesting and feeding grounds. They may spend time near the shore but are at home out in the big blue. And they are not the only creatures traversing the open ocean.

Great white sharks journey in substantial numbers from the Pacific coast of America to a seemingly unremarkable patch in the middle of the ocean, approximately halfway to Hawai'i. It was nicknamed the Café by researchers who discovered it because they originally thought its purpose might be to meet mating partners. In fact many juveniles, who are by definition pre-breeding age, also go there, so this reasoning is now thought unlikely. Once there, they appear to undertake deep dives much more frequently than a great white typically would nearer the coast, leading to the theory that they may come to this place to hunt squid. As yet we simply don't know for sure. A fascinating question is how long the great whites have been visiting the Café. The great white is an ancient species; one of evolution's great success stories. The fossil record suggests that they or their close evolutionary ancestors would have been swimming in these waters for millions of years. So have they always travelled to this spot? Did some change in the ocean conditions or aggregation of prey suddenly make it attractive to the super-sharp senses of great whites? Or was it one day discovered by an individual shark who regularly returned over its seventy-year lifespan, prompting others to follow? As we get small, intriguing glimpses of the lives of the great ocean travellers, we gather fragments of knowledge that both add to the picture we are assembling, of life in the ocean, and reveal that the canvas is far wider than we originally thought.

Grey whales migrate 8,000 kilometres each way from their feeding grounds in the Bering Sea to Baja California in Mexico, while humpback whales cover astonishing distances from tropical breeding grounds to the Southern Ocean and back again. One humpback was recorded making a round trip of 18,942 kilometres over 265 days. Tracking technology, such as satellite tags and acoustic tags, has become more reliable, smaller and cheaper, leading to a huge increase in its use and, as a consequence, a far richer understanding of the movements of many kinds of marine animals across time and space. The data it is yielding is finally giving us a window into daily life in the open ocean – in particular revealing the importance of certain areas as feeding grounds, nurseries and migration corridors.

Large-scale monitoring projects such as Tagging of Pacific Predators, which tracked over 2,000 animals in the

Pacific – from elephant seals to albatross – have revealed a complex variety of long-distance journeys that offer us a new way of considering the open ocean. At any one time a part of the ocean might appear completely devoid of life, but over the course of a year that same area could be frequented by a myriad of species perhaps attracted by plankton blooms, upwellings or activities far below the surface we do not yet understand. Or perhaps they are simply passing through. One of the lead scientists from the project, Barbara Block, has an evocative comparison:

> It has been like looking across the African savanna and trying to figure out where the watering holes are that a zebra or cheetah might frequent. Where are the fertile valleys? Where are the deserts that animals might avoid, and the migratory corridors species such as wildebeest use to travel from place to place?

We don't yet know the answers to these questions, but we are starting to build up a picture of how migrating species see the open ocean, and it is clear there are favoured routes, important locations and apparent reasons why different species move in different ways across this vast expanse.

Over recent years tagging projects have tracked sharks making enormous migrations, crossing the ocean in near perfect straight lines. Whales have been recorded making the longest purposeful journeys of any mammal on our planet. And bluefin tuna have been found to travel the entire Pacific shortly after their first birthday. We have only just discovered clues as to how these predators find their way out in the ocean far from land. A new, far more connected picture is emerging of the big blue.

Exactly how these great journeys are navigated is still being uncovered, but studies have shown that species such as green turtle and shark can detect the Earth's magnetic field and some seabirds appear to use the sun or the night sky to find their way. It may be that different species use a range of senses just as humans do – perhaps utilising Earth's magnetic field for general directions and physical features like ridges, seamounts and ocean currents, together with hearing and smell for specific localised navigation.

A standard map of the world, such as one you may find on the wall of a school library, might show the snow-capped mountain ranges of the Himalayas, Andes and Alps as white, the vast tropical forests of central Africa or South America in a dark green, the Kalahari and central Australia would be beige, the great Eurasian steppe a lightish green and Antarctica a lozenge of pure white along the bottom. The remaining 70 per cent of our planet would typically be indicated by solid blue. Occasionally the coastal and continental shelf waters – about one-third of the ocean – might be a lighter blue with the remaining open ocean darker. Of course, there exist much more detailed maps which do include some ocean features – canyons, ridges, seamounts and the like – but much of the time it is easy to portray the open ocean as a uniform blanket of blue, its structures and complexities out of sight and out of mind. Yet below the surface it is far from uniform. There are great navigation highways, nutrient upwellings creating hot spots of animal activity to rival any near shore, and gyres driven by Earth's rotation and winds all causing enormous volumes of water to circulate.

The open ocean can be described as the habitat beyond the continental shelf. To distinguish it from the deep ocean, which we have described earlier, we are concerned here with the portion within the photic zone, that is, the shallowest 200 metres of water. Below 200 metres little light penetrates and the potential for photosynthesis wanes, marking the beginning of the so-called deep. Spanning two-thirds of the ocean's surface, the open ocean accounts for half the surface of our planet. Much of it is classified as international waters, in other words beyond national jurisdictions, and as such there is very little legislation or policing. But while we still know comparatively little about the open ocean, in other ways our presence is ubiquitous.

Plastics now permeate every level of the marine food web. They have been found in the ocean's deepest trench and at great depths in all five ocean basins. A study in 2023 found that over 'the last two decades the number of marine species known to have ingested or become entangled in debris (of which the majority is plastic) has more than trebled'. It is widely understood that, when directly eaten – known as primary ingestion – plastic can have an impact on marine species. While it can often be hard to say for certain whether or not a particular animal was directly killed by plastic, we do know that ingestion is widespread. Sharp pieces of plastic can injure or pierce the insides of an animal and fragments can build up inside its body, blocking the gut. Of 55 sea turtle individuals (green, logger-head, olive ridley and leatherback) inspected by scientists after they were killed as by-catch from long-line fisheries in the Pacific, 50 were found to have plastics in their gut. In the Mediterranean, scientists found plastics in one out of every five of the swordfish, bluefin tuna and albacore tuna they analysed. And sperm whales appear to be particularly prone to plastic ingestion, as illustrated in 2016 when 9 out of 22 stranded whales were found to have items such as netting, rope, foil, packaging and even car parts in their stomachs.

Secondary ingestion, or eating prey that has eaten plastics, is also thought to be widespread. However, it is hard to quantify its impact as, for the most part, this concerns microplastics. These tiny pieces of plastic, usually defined as being less than 5 millimetres long, are now dispersed throughout our ocean. Some microplastics are manufactured deliberately for use in beauty products or toothpaste whereas others result from larger pieces of plastic debris degrading into smaller and smaller fragments. Those animals higher up the food chain are thought to be most affected. For example, seals eat large quantities of fish and over their lifetimes the microplastics from each of those fish may build up in the seals' bodies. Similarly, a study in New Zealand found that the plankton eaten by baleen whales contained microplastics. It was estimated that a mouthful of plankton contained around 24,000 fragments of microplastics, so over the span of a single day these whales could be ingesting 3.4 million of them. While some of these microplastics pass through their bodies, others accumulate in their organs, exposing the whales to the chemicals of which the plastics are composed.

There is another way plastic is harming marine species. Fishing gear snagged on a seamount, long lines accidentally cut at sea, damaged nets thrown overboard; there are many ways fishing equipment can end up adrift in the ocean or unrecoverable on the ocean floor. 'Abandoned, lost and discarded fishing gear', as it is now known, brings a dual problem. Most of it is made from plastic so it inevitably adds to the pollution in the ocean, but perhaps more insidious is the fact that it can often carry on catching and killing marine species long after any human is going to haul in that catch.

So-called 'ghost nets' drift for years in the ocean currents, entangling turtles and sharks; lost lobster pots continue trapping crustaceans even when their owners will never again collect them; and gear washed towards shore can smother areas of coral reefs or entangle seals and birds. In a self-perpetuating cycle animals entangled in fishing gear attract others that also become entangled as, by design, fishing gear is created to be hard for marine life to escape. This is extremely useful for fishers catching food for their community, but it is a horrifying waste of life when industrial-scale equipment endlessly floats though the ocean, snaring hundreds of animals no one will ever eat.

When we think of fishing, many of us picture brave men and women heading to big seas in small boats relatively near to shore to catch food for their communities. It is a difficult, dangerous and noble occupation as old as human

A group of mobula rays gathering around Princess Alice Bank, a remote seamount 30 nautical miles south of Pico Island in the Azores.

A blue whale mother and calf, off the coast of Loretto in the Gulf of California, Mexico.

A mixed baitball filmed off the Azores circled by Atlantic spotted dolphin and yellowfin tuna, with shearwater birds diving from above.

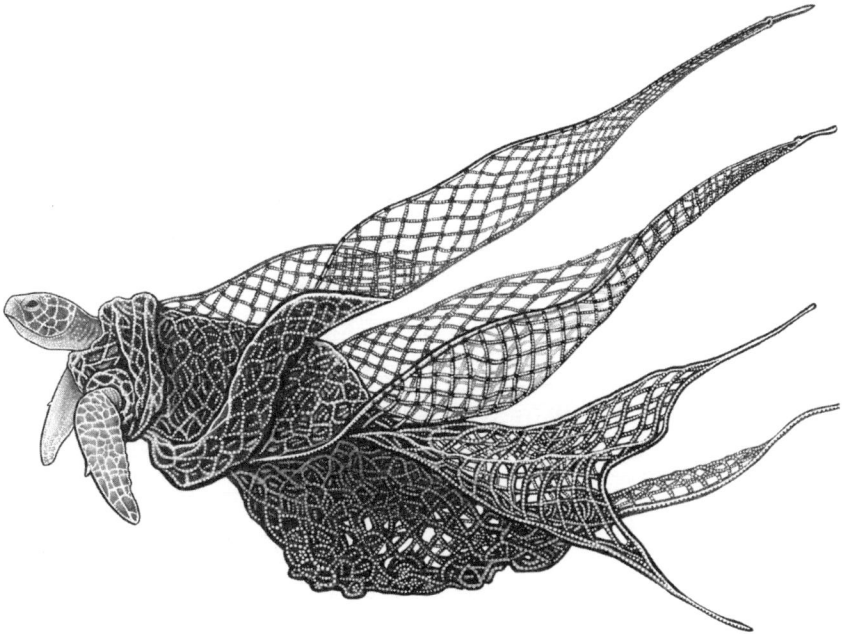

Ghost nets are aptly named, as they haunt the ocean for many decades after their original use has expired.

civilisation. They are characters from literature or children's stories, their boats those we see at small harbours, their families central to the survival of coastal communities the world over. But there is also a very different type of fishing that is so removed from the traditional occupation it probably requires a different name. Super-trawlers – giant factory ships; huge boats over 100 metres long with kilometre-wide nets and on-board processing facilities and freezers that enable them to travel far into the open ocean for months at a time. Equipped with high-tech kit for detecting, chasing and catching fish at greater distances and depths than ever before, the factory fleets can go anywhere and hunt anything. They are now even capable of catching tiny species such as krill in vast nets. This is fishing on a truly industrial scale.

Off the coast of Argentina, just outside their national waters, a glowing floating city can clearly be seen in night-time satellite images. It looks like a string of small islands each with a powerful lighthouse but is, in fact, hundreds of large foreign fishing boats stationed for months at a time. As night falls they turn on powerful lights to attract first plankton and then their real target – squid. Beyond national jurisdiction there is no limit to fish-catch numbers, no policing and no need to respect fishing seasons. No one knows accurately how much they are catching, what levels of by-catch, or other species, are being taken, or even what labour conditions the crews work under. But we do know that the squid they target are an important part of this ecosystem and food for many other species, from whale and dolphin to seabird.

As well as the fish – the factory fleet's intended target which do at least get consumed, albeit sometimes in the form of pet food or 'health' supplements – there is an immense amount of by-catch. By-catch is a broad term that includes everything caught in the nets, from the fish that weren't targeted but swim in shoals alongside those the factory ships were pursuing, to sharks, whales, dolphins, seals, turtles and seabirds. Some of this by-catch does end up being sold and consumed, but much is discarded. It is notoriously hard to monitor the behaviour of boats hundreds of kilometres from land in waters owned by no one; however, a 2019 study that collated data on by-catch from different fisheries around the world calculated that purse seine fishing (a curtain of net used to encircle a school of fish) in the high seas generated over 1 million tonnes of by-catch, midwater trawling almost 1 million tonnes and longline 400,000 tonnes.

But there is potentially cause for hope. While the open ocean is a global commons – supposedly equally there for all – an article in the international journal *Science* revealed that 77 per cent of global high-seas fishing vessels come from just six countries: China, Taiwan, Japan, Indonesia, Spain and South Korea. Those governments are estimated to be subsidising their fleets by 4.2 billion dollars; if the subsidies were removed then many of the operations would be unprofitable. In other words, the industrial fishing of the open ocean is not vital to the global economy – or even to the economy of any single nation. If stopped, the economic impact would be minimal and the recovery of our ocean substantial. There is compelling reason for change.

In 2023, under the United Nations Convention on the Law of the Sea, and after decades of discussions and negotiations, the High Seas Treaty was agreed. This enables the establishment of marine protected areas in international waters, requires an environmental impact assessment for activities in the open ocean, and demands an equitable sharing of the benefits from genetic resources such as medicines derived from species found in the high seas. It ambitiously sets out to make the open ocean easier to protect and consequently more likely to recover. But, as with all global legislation, there are political complexities:

it will be difficult and slow to reach agreement on proposed protections even once the treaty is ratified. Others have written at length about the politics of the High Seas Treaty, so instead we will focus on the biological and ecological impacts of protecting the half of our planet's surface that is open ocean.

Assuming we are unable to declare the entire open ocean a marine protected area, where should we choose to protect? As on land, there are some permanent features in the ocean, such as seamounts, that are clear hot spots for life and disproportionately valuable, but elsewhere it is less straightforward. Elephant seals, for example, were shown through the Tagging of Pacific Predators programme to be drawn to the meeting point between two large gyres – rotating currents which move and swirl in a manner governed by a combination of Earth's rotation and surface winds. These gyres connect in the Pacific where the warm waters of the southern subtropical gyre meet the cold, nutrient-rich waters of the subpolar gyre. The confluence fuels huge blooms of phytoplankton. The elephant seals and many other species benefit from the food along this gyre boundary, which suggests it would make an obvious and important location to protect; however, the blooms of phytoplankton and the feast they fuel move seasonally by hundreds of kilometres, and this creates a complication. Perhaps, then, the solution to high-seas protection is to view it on a scale commensurate to the open ocean itself, and create vast marine-protected areas.

This is exactly the hope for the Costa Rica Thermal

Dome. Despite its name, this dome extends along the west coast of Central America from southern Mexico to Guatemala, El Salvador, Nicaragua and Costa Rica itself. Some years up to 1,000 kilometres in diameter, it forms when warm coastal waters interact with wind and ocean currents to create an upwelling which brings deep, cold, nutrient-rich waters to the surface in the dome-like shape after which it is named. These nutrients create an immense biodiversity hot spot, and the resulting marine life is among the most spectacular and diverse on Earth. Blue whales, tuna, marlin, manta rays and many species of shark, turtle and seabird are regularly sighted. It is of significant economic value to neighbouring countries for tourism and both artisanal and sport fishing, but is also targeted by the fleets of factory ships. Although the dome continually changes in size it is essentially a permanent feature. Until now the issue has been that while some of the dome is within the waters of the Central American countries nearby and therefore within their jurisdictions, much of it is in the high seas and beyond the law. Many now hope that ratification of the High Seas Treaty will present the opportunity to protect the entire dome.

While safeguarding an area as large as the Costa Rica Thermal Dome will undoubtedly be of great benefit to marine life, species such as the blue whale might well have travelled thousands of miles to reach it and have encountered many dangers. A 2022 report by WWF titled *Protecting Blue Corridors* collated the research of over fifty teams tracking more than 1,000 satellite-tagged whales from a variety of different species and populations. When mapped out, the threats are very clearly illustrated. There are distinct migration routes for many whale species and all too often they cross shipping lanes, hot spots of industrial fishing and areas where currents create convergence zones in which large quantities of plastic, fishing gear and other rubbish accumulate. The report also suggests locations where noise and industrial development may be disturbing the whales' navigation and communication capabilities. It wisely proposes blue corridors to protect the whales' migration routes.

Increased knowledge of these routes and biodiversity hot spots like the Costa Rica Thermal Dome give us an indication of parts of the open ocean that are most vital to protect. But, as is often the case, climate change makes things somewhat more complicated. While the depths of the open ocean tend to maintain a stable mean temperature anywhere in the world unless disrupted by a specific input such as thermal vents or cold seeps, surface waters vary dramatically. Temperature has always influenced the species that live, breed or feed in water, but that once dependable constant is changing – and changing fast.

As we heat our world, the ocean is getting warmer too. The increased energy now in our global climate system is driving larger and more unpredictable storms, especially over the seas. Changes at the poles, particularly the Arctic, are combining with an overall warmer ocean, changing winds and altering ocean currents. It is hard to predict how this will end, but what we can say for sure is that hot spots, feeding grounds and migration routes in the open ocean will change in the future, and that makes it even harder to select which areas to protect and gives us another compelling reason to declare large areas of the open ocean a sanctuary for marine life.

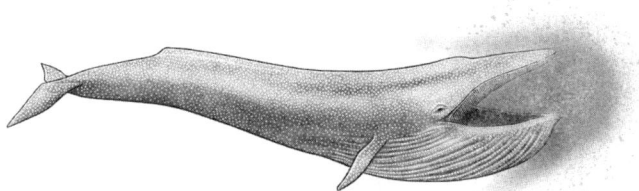

RETURN OF THE WHALES

Even fifty years later, and knowing how the story ends, the grainy, jumpy footage still chills you. In 1975 *CBS Evening News* aired footage of a small inflatable boat driving in front of a massive ship. Ironically, for a moment it looks as though this tiny vessel is riding the bow wave of the bigger one in much the same manner as dolphins often do – for fun, as far as we can tell. Then you notice the harpoon. The *Dalniy Vostok* was hunting whales and the Greenpeace activists, in little more than a dinghy, were putting themselves between the hunters and where they expected a whale to surface next.

It is hard to imagine the courage it must have taken to drive a small boat between a pod of scared leviathans and their pursuers, but from their subsequent testimonies the activists didn't anticipate that they faced a further danger – they hadn't believed the whalers would shoot their harpoons while they were directly in harm's way.

Then it happens: a whale suddenly surfaces just in front of the Greenpeace dinghy and the *Vostok* immediately fires. A harpoon, much larger than one would expect, shoots down from the bridge of the whaling ship. The missile flies over the heads of Robert Hunter and Paul Watson and slams into the flank of the whale. Shortly after, another harpoon penetrates a second whale's head.

The *Vostok* had unwittingly created the impetus for one of the greatest environmental achievements of all time and a shift in human consciousness that would reverberate far beyond whaling. There had been scientific concerns about the impact of commercial hunting on whale numbers for decades, but this footage, broadcast on mainstream evening news channels, is often credited as one of the two key moments when scientific concern was joined by public outcry.

The other was a surprising hit record by an American biologist. Roger Payne released his 34-minute *Songs of the Humpback Whale* in 1970. At the time you could have been forgiven for mistaking this title for any other seventies album, but his record name was quite literal. Roger specialised in bioacoustics and had stumbled across recordings of whale song through a chance encounter with a US naval engineer who had been using hydrophones to search for Soviet submarines and regularly located singing humpbacks instead. Enthralled by what he heard, Payne set out to make his own recordings of the humpback's songs, to play to audiences in schools and churches – eventually they were even played to the United Nations. He hoped they would be a powerful means to convey the whale's intelligence and individuality to a public largely unaware that more than 90 per cent of the global population of humpbacks had already been slaughtered.

Shortly before his death in 2023, aged eighty-eight, Payne recalled the impact of these recordings on live audiences: 'For the first 30 seconds there is mumbling and giggling as the audiences get used to the deep rumbling groans and high-pitched squeaks. But leave it longer and the audience would go totally silent as if in a trance.' The album sold more than 125,000 copies before National Geographic Magazine gave away a further 10.5 million.

The brave activists and the unlikely hit-maker secured two very different breaches in the public consciousness, persuading millions that whales were sentient, sensitive creatures and that commercial whaling was both brutal and unnecessary. The scale of the public response created an opportunity for both the environmental movement and the scientific community to attempt a new form of conservation. By the 1970s the world was very familiar with setting up national parks to conserve important ecosystems on land. Yellowstone was already over 100 years old and there is some historical evidence that Bogd Khan in Mongolia had been a protected site for over 200 years. The World Wildlife Fund had by this time been working to save endangered species for almost fifteen years, and scientific research organisations and universities had made huge strides in their abilities to monitor the health of both species and entire ecosystems.

But whales were an altogether different proposition. While there are examples of resident whale populations that could each be described as the 'responsibility' of a particular country, most members of most whale species migrate thousands of kilometres each year, crossing the waters of multiple countries as well as the shared space of the open ocean. As far as the law was concerned, for most of their lives whales were in common waters and were to be treated as little more than a common resource. Over the course of the twentieth century at least 650,000 fin whales were killed, as were over 4 out of every 5 humpbacks, while blue whales were pushed to the very edge of extinction, with just 4,700 left in the world by 2001 compared to nearly 350,000 a hundred years earlier. The 1960s was the peak decade for whaling during which time 700,000 animals were killed. As some of these figures are both historical and obtained from reported whaling catch, they should be considered an estimate – in all likelihood an underestimation – of the total number of whales killed.

It would be unfair to judge the whalers of the 1800s by modern standards. Doubtless many of today's 'normal' practices will appear barbaric to future generations, but while using whale oil for essential lighting in the early 1800s is one thing, by the late twentieth century there were few who would have described commercial whaling as vital to national economies or daily personal life. It is important, however, to reflect on the important difference between commercial and subsistence – or culturally significant – whaling. By the 1970s commercial whaling was an industry of large, powerful ships, like the *Dalniy Vostok*, that killed thousands of whales to make products which by this time could be replaced with other materials. This is quite different from traditional communities harvesting small numbers of whales either for their own subsistence or to continue important cultural traditions – practices that have endured for centuries without critically damaging whale populations.

There were modest attempts at regulating the commercial whaling industry through the International Whaling Commission (IWC), but they had little impact at this point. Campaign groups, lawyers, scientists and schoolchildren across the world demanded more. Yet despite the public outcry a quandary remained as to how to write a law to protect a sub-order of marine mammals in a part of the ocean that belonged to no one.

Humanity has been puzzling over the governance of the open ocean for centuries. Its history now reads like a proxy struggle for the debates, conflicts and aspirations of the last 400 years of the global human story. The first record comes from a late sixteenth-century proposition that a country should only control the stretch of sea adjacent to its coast that it might defend based on the reach of available weaponry. This became known as the cannon shot rule and in practical terms meant that anywhere more than 3 nautical miles from the shore was counted as international waters. Against this backdrop, fishing, exploration, war and piracy were enacted.

The world wars of the first half of the twentieth century brought rights over the open ocean to the fore once more. Industrialised nations now had the technology to fish, pollute, lay communication cables or travel across any part of the open ocean that they wished. Many nations pushed for greater control over the seas they were closest to. In a world of fossil fuels, battleships and fishing fleets, 3 nautical miles no longer seemed adequate. It was America that forced the issue. In 1945 President Truman, under pressure to secure valuable oil fields, claimed the United States' right over all natural resources on its continental shelf – some 200 nautical miles out to sea. While the action was principally taken over oil, the ruling also applied to fish and indeed whales. Other countries made their own assertions based on what felt valuable and appropriate given their particular location.

By this point humanity had at its disposal technology that allowed marine exploitation in a manner and scale unimaginable to the previous generation of lawmakers. Although the United Nations had spent several decades grappling with how rights in the ocean would be shared, it wasn't until 1982 that the United Nations Convention on the Law of the Sea was agreed. It accomplished many things but importantly did not resolve a way to manage marine life in areas beyond each country's waters. From the perspective of whales or fish, the open ocean remained lawless.

Complicated global legislation takes a long time, and whales did not have much left.

Looking back, it seems incongruous that we were travelling to the moon and hunting whales for oil during the same decade. But increasingly powerful ships were indeed pursuing whales in ever more remote locations, applying rapid advances in technology to target their dwindling populations across the international waters of the open ocean. Whales were now being used for products ranging from high-grade oil in missiles to pet food. But few could argue that the global industrialised economies of the 1970s were unable to find other sources to replace whale-derived ingredients. And it was perhaps this that finally tipped the balance in the whales' favour.

A flurry of legislation ensued, albeit over a decade. The United Nations recommended a ten-year cessation to commercial whaling, and the Convention on International Trade in Endangered Species of Wild Fauna and Flora warned that many whale species were on the verge of extinction, which added indisputable evidence to reinforce the public pressure and forced the IWC to vote for a moratorium on commercial whaling in 1982. Many feared this was too late and that whale populations might already be too low to recover. But slowly the regenerative power of the ocean began to yield hope. And as whale numbers gradually increased, scientists watched their recovery and discovered more about their fascinating lives and role in the ocean ecosystem.

Off the coast of Dominica a female sperm whale leaves her calf at the surface and dives. The deep-water trenches of this region are perfect for her purpose. She descends a kilometre with ease, her heart rate slowing to just five beats per minute. Comfortably able to reach three times this depth and remain there for forty-five minutes, the sperm whale is magnificently adapted to hunt far below where even the faintest traces of sunlight can penetrate.

Our whale emits a series of fast-paced clicks so powerful that if she was in shallow water and you were swimming nearby, you would feel them deep in your bones. Echolocating her prey, she speeds up, moving from surveyor to raider in an instant. The largest of the toothed whales and therefore the biggest predator on the planet, she is a formidable hunter pursuing a dangerous prey.

The colossal squid fires a cloud of ink from the sac in its mantle. The absence of light means the ink can't be intended to blind its pursuer in the traditional manner of its more surface-dwelling cousins, so scientists suspect it might use luminescent ink to distract the whale. The whale closes in and grabs the squid in its teeth, but this hunt is far from over. The squid's eight arms and two tentacles are lined with suckers and hooks. The tentacles are over 2 metres long and in an instant the sperm whale's head is engulfed in a thrashing, writhing entanglement of vicious limbs. The whale bites hard. She is still nursing and needs a lot of food to produce the vast quantity of calorific milk her calf requires daily. The squid has three hearts and they are now pumping the squid's blue blood into the ocean through multiple lacerations. The struggle continues but the squid is losing strength. Its suckers and hooks no longer have the power to damage the whale, and its blood loss is so significant it begins to lose consciousness. The whale swallows the colossal squid, beak and all, and heads back to its calf over a kilometre above, in the sunlit Caribbean.

It is a battle every wildlife filmmaker would dearly love to record – surely the greatest of all hunt sequences, for now only visible in our imaginations, pieced together from the scars on sperm whales' heads and evidence found in their stomachs. It was the beaks that gave it away. Colossal squid have a large, hard beak formed of a substance called chitin and surrounded by muscular tissue. Squid have a narrow oesophagus, and the beak chops food into smaller chunks to allow it to pass into the digestive system. When eaten by a sperm whale, the squid's tissue quickly dissolves but the beak remains. Scientists dissecting the bodies of stranded sperm whales routinely find large numbers of these beaks in a dead whale's stomach.

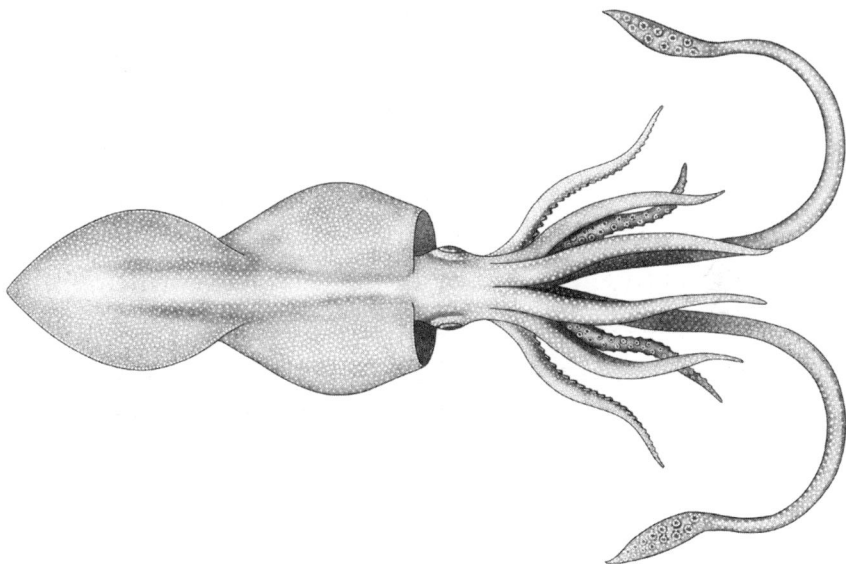

The colossal squid: a fearsome predator in its own right, and presumably not an easy catch for a sperm whale.

Strangely, it turns out that humans have unknowingly been collecting beaks that have passed through sperm whales for at least a thousand years and using them in perfume! Ambergris has been called 'floating gold'. Washed up on beaches, it can look a little like a rock, its colour varying from dark black to fairly light as it bleaches over time in the sun, and it releases a musky odour when dried. For hundreds of years perfumers extracted an alcohol called ambrein from the ambergris and used it to prolong the scent of perfume. But no one knew what it was or where it came from. Eventually in the 1800s the truth was revealed. On rare occasions the indigestible beaks of squid and other cephalopods accidentally move into the sperm whale's intestines and bind together in a clump. Scientists believe the ambergris is formed to protect the whale's intestines and organs from the sharp beaks. How this leaves the whale's body is still a matter of some dispute, but ambergris was a highly sought-after treasure and, despite available synthetic alternatives, it is still used in a few expensive perfumes today.

Fascinating as this mystery was, the digestive habits of the sperm whale revealed a far more important discovery, one with far-reaching consequences for life in the ocean and for each of us. Sperm whales feed in the depths but defecate at the surface. Whale poo contains both iron and nitrogen, which fertilise the sunlit surface waters. These elements combined with the sun's energy fuel blooms of phytoplankton. This process has a dual benefit. It brings otherwise lost nutrients from the depths to surface waters, stimulating the microscopic plants which underpin entire ocean food chains, thereby giving benefit to countless marine species. Additionally, the vast blooms of phytoplankton draw down huge amounts of carbon dioxide from the atmosphere as they grow, helping to slow climate change. Studies suggest that phytoplankton not only contribute up to half of all the oxygen in our atmosphere but they also capture an estimated 40 per cent of all the additional CO_2 we have produced. Of course, not all this is attributable to whale poo, but the International Monetary Fund calculated that even a 1 per cent increase in phytoplankton due to whales would be the equivalent of the instant appearance on Earth of 2 billion mature trees. So, if one day whales were to return to their total pre-whaling numbers of between 4 and 5 million (from about 1.5 million today), the impact in terms of both ocean productivity and climate change would be profound.

There is great cause for hope. Since the moratorium whale numbers have been increasing. Some populations are recovering fast. Others, such as blue whales, are only slowly creeping up. But the signs are positive. In theory we would expect that as long as the moratorium on commercial hunting remains then most whale species should continue to recover, albeit slowly. But we shouldn't be complacent. While there is justifiable concern about some nations pushing to restart commercial whaling or continuing it under the guise of 'scientific whaling', other risks are equally concerning but perhaps less widely known.

Each winter 10,000 North Pacific humpback whales – approximately half the entire population – travel to the Hawaiian Islands, primarily to breed. They journey from

133

higher-latitude feeding grounds such as British Columbia and the Gulf of Alaska. Along the way they cross the open ocean, passing through major shipping lanes and the Great Pacific Garbage Patch. Almost half the garbage patch is discarded fishing gear, which can entangle the whales, killing or debilitating them; ship strikes are a common occurrence here too.

Up until only recently the open ocean was considered so vast that human activities could never affect it. Governments routinely dumped toxic waste and even chemical weaponry at sea, never imagining that it would damage food chains and potentially find its way back to our bodies. But much as we have discovered on land, while protecting individual species from hunting may have once been sufficient when there were still vast areas free from human impact, now – in the Anthropocene, the age of humans – we are everywhere. As such, saving a species isn't enough; we must look at the whole ecosystem.

A new High Seas Treaty was adopted by the United Nations in summer 2023. It was justifiably welcomed, albeit cautiously, by many conservationists as a first step to creating marine protected areas outside individual countries' own jurisdictions. It is perhaps possible that blue corridors and other proposals for high-seas protection will follow in due course, but it could take a lot of time for real change to occur. The treaty must be ratified by at least sixty member states, and before they can become real refuges for marine species, protected areas must first be proposed and then gain the support of regional fisheries organisations, who may argue to allow fishing to continue.

The reality is that global legislation, while vital, is a lengthy process. Because of this some are acting now to see if they can buy whales more space and time.

Dr Enric Sala stands on the cliffs of Dominica looking out on the newly designated sperm whale reserve. Dominica is home to at least 200 sperm whales and its deep canyons and seamounts are probably visited by many more. Unusually, scientists who have studied the whales believe the resident population really are just that – resident. They remain local for much of the year, appearing to travel much smaller distances than most whales, sticking to the eastern Caribbean. Enric's programme, National Geographic Pristine Seas, works with national governments and local communities to survey their marine life and help establish science-based marine protection. His team have worked closely with the government to establish the Dominica Sperm Whale Reserve.

Dr Shane Gero, who has been studying Dominica's sperm whales for over twenty years, notes that sperm whales off Dominica have been declining steadily over that period. Across the ocean, including in the eastern Caribbean, whales are hit by ships, become entangled in fishing gear and are affected by pollution. But in Dominica that could all be about to change. The country has declared the resident sperm whales to be national citizens: after all, as Enric Sala notes, 'their ancestors were likely there before humans'.

To protect the sperm whales and allow for their recovery, Dominica has designated an area 900 square kilometres in size – larger than the island itself – where commercial shipping traffic will be prohibited except for a channel to the main port. The sperm whales now have a safe refuge to feed, rest and raise their calves.

The Dominica government has been inspired by the success of protecting mountain gorillas in Rwanda and Uganda, where the tourism revenue and jobs that have been created far outstrip any potential income from clearing the forest, and the conservation of the mountain gorillas is now regarded as one of the best examples of ecotourism in the world. When announcing the sanctuary in 2023, Dominica's then prime minister Roosevelt Skerrit highlighted twin benefits for his country. Limited numbers of high-paying tourists could snorkel with the sperm whales or take whale-watching boats, and the revenue from these activities would sustain the monitoring and policing of the reserve, fund education programmes and provide the community with a range of other benefits. Additionally, the sperm whale's feeding at depth and defecating at the surface would fuel the growth of plankton which both supports a healthy marine ecosystem and draws down carbon dioxide from the atmosphere.

Sala adds a third benefit: the protected area will give other marine species a chance to recover, replenishing fish stocks and leading to a more sustainable long-term artisanal fishing community.

It is a perfect example of a positive tipping point – a change that will keep creating ever greater self-perpetuating improvements. By protecting the whales you create a marine sanctuary. Inside that sanctuary other species also recover and thrive. The whales bring nutrients from the deep, fertilising the surface waters and accelerating the recovery of marine life. As the marine life becomes ever more spectacular, the tourist revenue grows. As the sanctuary thrives, fish populations get so big they grow and spill over outside the reserve improving catch for local communities. The sanctuary gets more and more valuable to Dominica.

Ending commercial whaling was one of the greatest achievements in the history of conservation, and is rightly still celebrated fifty years later. As many whale species recover, their renewed presence reveals how valuable they are to the ocean and to us. But in the Anthropocene, when the human world touches every part of the ocean, if we truly want to save the whales then we must now protect their home.

4

KELP FOREST

SEA OTTER, SOUTHERN CALIFORNIA

Drifting in a wetsuit above a submarine forest in southern California, I found myself alongside one of the most blissful of creatures. On its back, all four paws tucked into its body fur for warmth, gently rolling in a manner that brought to mind a swaddled newborn baby, lay a southern sea otter. They were once seen linked together in rafts hundreds strong, but this one was alone and it seemed to be quite unconcerned by my comparatively clumsy attempts to float nearby.

I was no more than 200 metres offshore, preparing to record a piece to camera on the wildlife of the Pacific coast of North America. If I looked towards shore I could still see a few houses and the odd car, yet if I looked down I felt I was in a wilderness. The forest beneath my otter companion and me was one of giant kelp, each frond anchored by a holdfast to a rock on the sea floor some 45 metres below. I only had a snorkel so those depths were out of reach. But not for my neighbour. Periodically it dived down beyond my sight. A sea otter's hind paws are fully webbed and reasonably flat, so although they are capable of moving fast on land, they are also extremely effective divers. They can close their nostrils and ears, and their lungs are so big that not only can they float without any effort but they can also remain underwater for around four minutes at a time. This otter was diving in one of the richest marine environments on the planet, so finding food was no problem.

These cold but sunlit waters are naturally rich with nutrients, but it is the kelp which makes them so biodiverse. The great fronds provide food and shelter for a multitude of different species. Lingcod, sheepshead and rockfish swim between the giant blades, which provide very effective protection for their fry. Snails, abalone and urchins eat the fast-growing blades. Octopus dart between rocks, jellyfish pulse through the kelp stipes and seals, rays and even an occasional whale also forage in this diverse ecosystem.

The sea otter suddenly reappeared beside me. It had used its sensitive whiskers and front paws to locate and collect a clam from the sea floor. Once back on the surface and floating on its back, it produced both a clam and a rock from a pouch of skin under its forelegs. I watched, captivated by the practised skill with which it balanced the rock on its belly and then smashed the clamshell repeatedly against it until the shell broke apart. Sea otters are one of the few species that, like human beings, regularly use tools. Even more remarkably, perhaps, than doing so while they float on the surface, they have also been observed deep underwater using the same sort of action to dislodge giant abalone from their shells.

They eat a wide range of the inhabitants of the kelp forest but one is of particular importance to them – sea urchins. These near spherical, spiny invertebrates live on the rocks and floor of most of the world's seas. In a healthy, balanced kelp forest they play a key role, acting like a kind of kelp gardener. They gnaw away at the algae growing on the rocks and in doing so create pits that enable the kelp to anchor their holdfast. From these the great blades soar upwards at the almost unbelievable speed of up to 60 centimetres a day.

Left unchecked, the urchins can destroy such a forest by eating the holdfasts that keep the kelp in place. As a consequence, the plants drift away to the open ocean and ultimately sink to the sea floor. If urchin numbers grow really big, the whole kelp forest disintegrates. When the ecosystem is in balance, this does not happen as there are several species, such as the Asian sheepshead wrasse and sunflower sea star, which eat sea urchins. But it is the sea otter that is the most effective urchin predator. An otter needs to eat almost 25 per cent of its bodyweight every day, so has a huge appetite. A single otter can eat hundreds of urchins a day, flipping them over to avoid the long spines and attacking them on the underside where the spines are short.

The importance of the sea otter was revealed when almost 200 years of hunting brought them to the verge of extinction. Unusually for a marine mammal, sea otters don't have blubber, so they were not targeted by humans in the way that seals were, for the extraction of oil and other products. In lieu of blubber, however, they have the thickest fur of any mammal, a double-layered pelt which enables them to keep warm in these frigid seas, and in the eighteenth and nineteenth centuries they were hunted for that in their thousands. As a consequence, the global population fell from 150,000–300,000 to less than 2,000 individuals.

Urchin numbers exploded, and as a result many kelp forests all but vanished, taking with them much of the other life that used the forests for food or shelter. The delicate balance of this complex system was devastatingly disrupted by the targeted removal of a single keystone species.

But promisingly, this process can also happen in reverse. Since hunting sea otters was banned in the early 1900s, numbers have slowly recovered across significant parts of their historic range. Recovery is not yet complete, but where it has happened the effects on the kelp forests are often spectacular. As the otters feast on the urchins, the kelp gets some respite. Being so fast-growing, it quickly begins to provide habitat that attracts other species, including other urchin eaters such as the sheepshead wrasse. In many places, including the bay in southern California in which I was swimming, the kelp is once again so thick at the surface that birds such as egrets can stand on it while resting or fishing. From a distance they seem to be walking on water – a fitting image, as the recovery of this kelp forest seemed little short of miraculous.

DA

It was an expedition originally planned to last two years. But there was so much to be explored that it was nearly four years after leaving Plymouth, England, when the ship finally made the landfall which ensured its place in history. The expedition had first sailed south across the Atlantic, through the Azores and Cape Verde; it had surveyed the coast of South America, rounded Cape Horn, and then, after a stop in Lima, travelled some 1,000 kilometres across the Pacific to arrive at an archipelago of nineteen islands. It was 15 September 1835, the ship was HMS *Beagle*, and Charles Darwin wrote a note describing the Galapagos as 'bleak'.

He did, however, also note the giant kelp forests encircling many of the islands. In time, of course, this passing observation would be eclipsed by his world-changing study of Galapagos finches and the evidence they provided for his theory of evolution by natural selection. But, throughout the five-year voyage of the *Beagle*, Darwin regularly observed a variety of kelp species in cold, clear waters. It clearly fascinated him to such a degree that he concluded that kelp was 'one marine production which from its importance is worthy of a particular history'. He added:

I know of few things more surprising than to see this plant growing and flourishing amidst those great breakers of the western ocean. The number of living creatures of all orders, whose existence intimately depends on the kelp, is wonderful. A great volume might be written, describing the inhabitants of one of these beds of sea-weed . . .

Like many scientists after him, Darwin found that the closer he looked at the kelp forest the more life he realised it supported, remarking that each time he shook a mass of kelp he would discover 'a small pile of fish, shells, cuttlefish, crabs, sea eggs, starfish and crawling nereidous animals of a multitude of forms'. Comparing the diversity of life within the kelp forests to that of the tropical rainforests of South America he had visited earlier in the expedition, he suspected that the kelp forests might be of even greater importance:

> *If [tropical rainforests] should be destroyed in any country, I do not believe nearly so many species of animals would perish, as, under similar circumstances, would happen with the kelp. Amidst the leaves of this plant numerous species of fish live, which nowhere else would find food and shelter: with their destruction the many cormorants, divers, and other fishing birds, the otters, seals and porpoises, would soon perish also.*

When many of us think of algae we probably conjure up an image of spreading greenish growth, perhaps around a pond or carpeting a riverbank. But imagine algae 50 metres tall, able to grow up to 60 centimetres a day, in huge forests which thrive within cold, clear waters that enable them to photosynthesise efficiently yet are turbulent enough to deliver regular nutrients. Kelp is algae on a supercharged scale. While resembling a terrestrial forest in some ways, it has no roots in the ground and therefore no ability to draw nutrients from the soil. It attaches itself by a stalked

Halfmoon fish swimming in a giant kelp forest
off the Channel Islands, California.

A young California sea lion rests
in the canopy of a forest of
giant kelp, Santa Barbara Island.

A California sheephead eating purple urchin just off the shore of the Channel Islands. If the numbers of urchin-eating fish and crustaceans fall too low, the urchin population can explode and destroy a kelp forest.

stump called a holdfast to rocks, oyster beds or similar solid substrate on the ocean floor, and draws everything it needs to grow from the sunlight and the water.

This is giant kelp. It grows to a prodigious height, but there are 134 other species of kelp in our ocean and a further 550 species of forest-forming algae which often live alongside it but are not kelp. Terms can sometimes get a little confusing in this area as quite often seaweed and kelp are used interchangeably, but essentially kelp is a type of seaweed and both are a type of alga.

Kelp vary hugely in size and form. Some grow along the seafloor, others are helped up in the water column by their stipe, while certain species rise even further to reach the surface of the ocean. You can find kelp with long flowing blades or bulbous bases: they may have densely packed bushy fronds or wispy thin ones. While some are huge, many species are only a couple of metres high – but even those can still grow to be over a decade old. It is this diversity that has enabled kelp forests to flourish along more than 30 per cent of the world's coastlines ranging from temperate to polar regions, covering an estimated area of more than 2 million square kilometres in nutrient-rich waters.

No matter how nutrient-rich the water may be, kelp requires ample sunlight to penetrate the water and fuel its growth. If the water is clear enough then kelp has been found to grow at depths of 45 metres in the seas off central California, at 68 metres in the Southern Ocean and at an astonishing 131 metres along an offshore bank in southern California. And wherever kelp grows, at whatever depth and to whatever height, life abounds.

A kelp curler, a small shrimp-like crustacean with eight pairs of limbs, chooses a kelp frond carefully. It pulls one edge of the blade over its body until it touches the other edge, creating a pod. It sticks the frond together using a substance it generates known as amphipod silk, and in doing so has very quickly created for itself a perfectly camouflaged shelter roughly the size and shape of a pea pod. It remains hidden within its shelter, but there is one complication: its home is also its food. Over four or five days it eats so much of the blade that it can no longer provide adequate shelter. The kelp curler must move. It uses its front antennae to sense the surroundings for danger, and when confident the coast is clear, it darts out. It hops quickly between fronds, searching for the right one but not wanting to risk being in the open for longer than necessary. And with good reason: kelp curlers are firmly on the menu for many of the fish in the kelp forest.

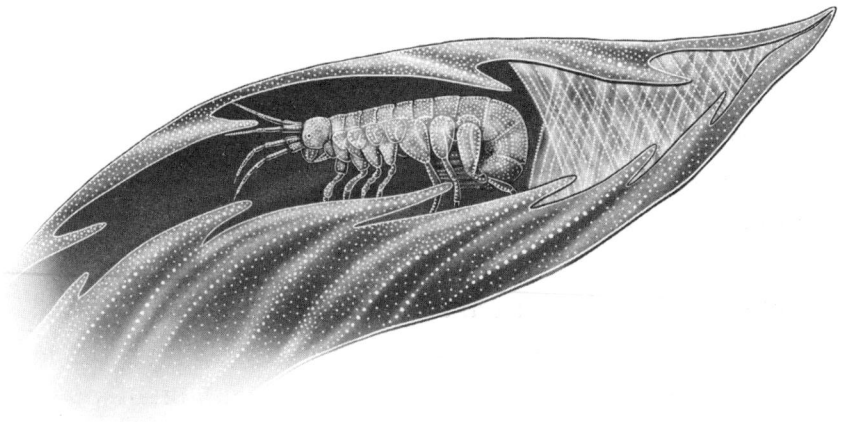

A kelp curler in its home/meal.

The kelp curler is just one example of countless connections between species within kelp forests. When swimming through one for the first time, your senses are overstimulated by the life around you. At the surface the fronds are often so thick and entangled that seabirds perch on them to rest or wait for a meal to pass by. As you dive below this dense upper mat there is a little more space and it is possible to weave between the 'trunks' of the forest. With time to pause, you can appreciate the similarities with the layers of life in a tropical rainforest. There is a community in the canopy, nearest the sunlight – they exploit the resources below but are quick to return to the light. Below the canopy there are creatures like snails and shrimp which spend much of their lives on a single frond. Midsized residents such as kelp bass and black rockfish are both predator and prey – they dart away instinctively as you swim towards them but don't flee entirely. Then there are larger animals like sea lions – clever, social and inquisitive, similar in that way to monkeys in a rainforest. Rarely preyed upon, they search the forest for food or fun – they approach confidently and swish away.

Then as light fades you reach the understorey, or sea floor. Lack of light doesn't hamper the richness of life down here. Detritus from above feeds lobsters, crabs, sea stars, urchins and molluscs such as seahares. The rocky sea floor is packed with invertebrates: a scooped handful of the coarse sand between the rocks could, and often does, contain microscopic species new to science. As in a rainforest, large predators are often sensed around the edges but not seen, generating a vague feeling of unease. It seems wise to return to the surface.

The sheer physical scale of a kelp forest also shapes the world around and within it. A large kelp forest will change the way water flows – this is often visible from the shore, with messy and rough water before the wave reaches the kelp forest and smoother clean waves afterwards. This aspect makes kelp coastlines much prized by surfers, but the smoother waves also have a practical benefit to the forest as they encourage nutrients to fall from the water column to nourish the forest dwellers below. The mass and physical presence of a kelp forest further shapes what lives there. Its scale, and therefore the scale of life it supports, attracts oceanic predators, and the shade it creates when its fronds reach the surface determines what can grow beneath. As it grows, kelp creates an entire habitat while also being the base of its food web.

Kelp forests may be extremely productive, but as with many of our planet's habitats they work best when part of a balanced, flourishing ecosystem. Studies from both San Francisco and South Africa raise a fascinating but concerning possibility that other changes in our ocean, seemingly unconnected to kelp, can alter the balance of life in a kelp forest. The forests of bull kelp around the Farallon Islands, off the coast of San Francisco, support populations of five different seal species. This abundance attracts top predators such as great white sharks – in fact some of the highest numbers in the world. More than 100 great white sharks have been known to aggregate in the area at certain times of the year, primarily to hunt the seal population. However, the supremacy of these sharks seems to be overruled once orcas enter the area. Orca are one of the few species that actively hunt and kill great white sharks, so when orcas ventured within a few kilometres of the Farallon Islands, the sharks moved away and fewer seals were attacked. A similar trend has occurred in South Africa. Scientists studying this phenomenon fear it may be leading to an imbalance of seals within the kelp forest as, for now, the orcas don't appear to be eating them at the same rate as the sharks once did. The increased numbers of seals may in future, therefore, reduce the fish and lobster populations, giving greater opportunities for urchins to multiply and overgraze the kelp. The food chain may have been knocked off balance.

There is much for us still to learn. What is driving the orcas to these areas? Is it a result of some other imbalance in our fast-changing ocean? Will the orcas in time keep the seal populations in check as the sharks once did? We do not yet know, but this example demonstrates the intricacies of kelp ecosystems and interdependencies of species that rarely interact directly with one another.

Kelp forests even create habitats hundreds of miles from where they themselves grow. In large storms kelp can become detached from the sea floor. This is a normal part of their life cycle; indeed many kelp species have evolved to release their spores at the moment when storms arise, to help disperse them far and wide, thus extending the forest. For many decades we assumed that most of the kelp that ripped loose in storms and swells was then washed up onshore and thereafter decomposed. This certainly does happen, as anyone who has walked along a beach in the temperate zones can attest, but as scientists have studied more of the ocean, they have realised that significant amounts are also washed out to the open ocean. Giant kelp has been recorded floating in large rafts hundreds of miles from shore, and over 150 different species of marine organisms have been recorded on or around them. One study even estimated that there may be 70 million kelp rafts afloat in the Southern Ocean at any one time.

Such is the volume of this floating kelp and the species carried with it that there is now active research into whether it could be responsible for transporting non-native, and potentially invasive, species across vast distances, and, if so, what role it may once have had in how species have colonised different parts of our world. It has been argued that the small clam *Gaimardia trapesina* spread across the Southern Ocean by hitch-hiking on kelp rafts floating in the powerful Antarctic Circumpolar Current. In theory, a kelp raft that stays intact for 100 days could travel 4,000 kilometres on such a strong current as that.

But most of these kelp rafts are not thought to wash up on faraway shores – they end their journey in the deep. As far as we can tell, they eventually sink due to a combination of storms, their air bladders degrading and losing buoyancy, or because of the sheer weight of other species growing on them. When they sink to the depths, they provide important food for deep-sea grazers and microbes, which break them down. Unless the deep-ocean bed is disturbed the carbon that the kelp once drew out of the atmosphere during photosynthesis may be sequestered thousands of metres down. In this way kelp forests greatly aid us in tackling climate change. But the kelp itself may be sensitive to our heating ocean.

It is well documented that most species of kelp only thrive in cooler water – we need just to look at some of the locations where they are found: the Atlantic coasts of Europe and North America, Chile, California, British Columbia, Southern Africa. While the ocean temperature of these areas rarely falls below 5 degrees Celsius, it is the summer temperature that appears to be most important, as it is heat not cold that is inimical to their growth. When marine heatwaves devastated kelp forests in Western Australia in 2011 and California in 2014, scientists began to worry that as we heat our world, we will limit the growth of kelp forests. This creates a negative climate feedback loop because kelp forests are key to carbon capture; if temperatures rise and less kelp grows then there will be even more carbon dioxide in the atmosphere, leading to increased heating.

In Diablo Cove, California, the construction of a nuclear power plant led to an important long-term study to test this hypothesis. The coolant water released from the power plant increased local water temperature by 3.4 degrees Celsius (a similar increase to an El Niño event). The scientists conducting the study recorded marked changes in the entire kelp forest community: there were drastic declines in several kelp species while other types of alga increased in abundance by 3,000 per cent, an explosion in urchins and grazing gastropods, and an increase in predatory species such as sheepshead wrasse and bat rays. It is an example of what can happen when cold water ecosystems heat up but also a warning that the responses are complex and difficult to predict. What it does clearly show, however, is that the changes can be substantial and disruptive, and they can happen very quickly.

Some have posited that while increasing ocean heatwaves will be damaging for existing kelp forests, our heating world is also allowing certain species to move closer to the poles. This may benefit some species of kelp that have hitherto been limited in how far north they can spread due to sea ice cover. There may be some truth to this possible silver lining to an otherwise dark cloud, as interestingly the kelp that already grows in the Arctic manages both to maximise growth potential during the long daylight months and, more surprisingly, survive the long dark months when photosynthesis is limited or impossible. How they do this we do not yet know. Of course, before we get too excited about new kelp forests in the far north of our planet, we should note that if kelp species succeed in spreading to higher latitudes, it may sometimes be at the expense of the other species that currently live there but could not endure the warming climate.

With all the changes occurring in our ocean the obvious question is: what is the current condition of the world's kelp forests? The answer is somewhat less clear than one might hope for a habitat typically found close to shore. Data, particularly historical data, is inconsistent: some areas are very well studied and monitored, others neglected. But broadly speaking we can say that kelp forests have declined around the world over the last hundred years and are still declining today. Understanding long-term changes in kelp forests is complex and scientists take pains to avoid overstating trends. But research does appear to show that, overall, kelp forests in Western Australia, New South Wales and parts of New Zealand are declining, as they are certainly doing in Maine and the wider north-eastern United States coast. By contrast, many in Greenland, Southern Chile and parts of Norway are stable or even increasing.

However, the picture can change fast. Kelp forest ecosystems are extremely dynamic. Large areas can be easily destroyed by storms but can also grow and recover quickly. The two most significant long-term threats are changing water conditions – in particular increased nutrients from pollution and ocean warming – and fishing. Overfishing can remove species that are otherwise important for maintaining balance in a kelp forest, such as urchin-eating lobsters and wrasse.

We know that the future of kelp is central to the future of our ocean and important for climate change. But it also benefits us in other ways. Kelp has been important to coastal communities for as long as humans have lived by temperate seas. For thousands of years it has been used as fertiliser, livestock feed and as an important part of our own diet. As early as the 1700s it was harvested in great quantities and used in chemical production. Across Ireland you can still find the remains of kelp kilns where the dried kelp was burnt to create an ash used in soap, glass, the dyeing of cloth and, much later, the production of iodine and gunpowder. Indeed, the word 'kelp' was originally used to describe the ashes of burnt seaweed; it was only later that the label was transferred to the species of large seaweed because they were the best for 'kelp' production.

Large-scale harvesting of wild kelp has a mixed reputation – some operations are regarded as well managed and sustainable while others face strong local opposition. But kelp's swift growth has also led to increased interest in ocean farming, and there are clear potential benefits. Kelp is what is known as a zero-input crop – in other words it will grow without fertiliser, pesticides or even much management at all. It could provide habitat for other marine species while it grows and, perhaps most attractively, it is a crop that can be grown without competing with either existing farming or nature on land. There are many products that can be made from it, which could perhaps then reduce our demands on the land: kelp is already being used on a small scale as fertiliser, but this could be scaled up to replace industrially produced chemical fertilisers; it is already being used as a source of algin – a binding agent for a myriad of products from cosmetics to toothpaste; and kelp is also of great interest to the pharmaceutical industry thanks to the variety of bioactive compounds it contains. There is even debate around the potential large-scale cultivation of kelp for harvesting, taking out to sea and deliberately sinking to the deep ocean as a means of capturing carbon – though this raises questions about the impact of growing huge amounts of biomass in one ecosystem and deliberately dumping it in another.

However, while the potential of kelp farming is an important avenue to be explored, we should be careful to avoid conflating that with the vital requirement for wild kelp forests. The former certainly seems worthy of exploration, whereas the latter is essential for a healthy ocean, to retain livelihoods for coastal communities, and for us to have a fighting chance of stabilising our heating world.

Kelp forests have long been underappreciated. Perhaps it is because their wonders are less immediately visible to us than those of a terrestrial forest or a coral reef, or perhaps it is because we have only recently truly understood their importance. Kelp is often overlooked in national marine management plans and was not even included in the targets for the 2021 to 2030 UN Decade on Ecosystem Restoration. Yet perhaps its time has come. As fish catch is falling for many communities and the reality of climate change hits us, we search for hope. Kelp makes for an unlikely silver bullet, but if someone told you there was a new wonder product that could fill your coastline with fish, draw down vast amounts of carbon from the atmosphere and provide a home for some of the most enchanting wildlife on the planet, you would be interested, wouldn't you?

OCEAN

THE GREAT STORM, SUSSEX, ENGLAND

A severe depression was sitting over the unseasonably warm waters of the Bay of Biscay. Formed by the collision of cold polar winds and warm tropical air, the dramatic temperature difference between the two forced the warm air to rise rapidly and a deep low-pressure zone to form – the perfect conditions for a very large storm. On 15 October 1987 it moved north, heading straight towards the UK and northern France.

Shoreham-by-Sea is a small costal town. The silt-rich waters of the River Adur meander through the open meadows and farmlands of an area known as the Weald, picking up the overflowing crystal-clear chalk stream running off the South Downs National Park and taking it out to sea between the town itself and the ever-moving shingle spit of Shoreham Beach. The people of Shoreham, like those of much of the UK, went to bed that night confident that the coming storm would mean extremely high rainfall but otherwise there was nothing to worry about.

At 3.10 a.m. an anemometer measuring wind speed on top of the Isle of Wight's iconic Needles cliffs was ripped from its mounting as winds of approximately 185 kilometres per hour tore towards Shoreham, whipping the sea of the English Channel into violent waves that smashed into the beach and the rest of England's south coast.

That morning residents of the UK awoke to the aftermath of what the Met Office called a 'once in 200 year event'. Living in a part of the world accustomed to the mild rainy weather of a maritime climate tempered by the warmth of the Gulf Stream, they were quite unprepared for what they saw. Roofs had been torn off, cars flipped over and power lines felled, leaving hundreds of thousands of homes without electricity. A combination of sodden soil and trees yet to drop their canopies of summer leaves created the perfect conditions for near hurricane winds gusting throughout the night to uproot over 15 million trees. Every part of the country could point to ancient, locally famous trees destroyed that night, but perhaps the most widely remembered was the town of Sevenoaks in Kent, in modern times known for the seven trees planted there to mark King Edward VII's coronation, where only one remained standing when morning came.

Understandably, in the weeks that followed most of the news focused on damage to property, deforestation and the tragic loss of eighteen lives. But in Shoreham there was another, more familiar problem. Huge piles of kelp uprooted by the relentless energy of the waves that night had been dumped on the beach; similar sights were reported all along the south coast. While local residents remarked upon the sheer volume of kelp, the sight itself was not unusual for just off the shore lay a vast, wondrous forest.

From the headland of Selsey Bill all the way to Shoreham, some 40 kilometres to the east, lay an abundant kelp forest teeming with life. Black sea bream swam through the fronds feeding on seaweed and the small invertebrates living on it; European sea bass searched for shrimp and small crabs; squid laid white finger-like capsules of eggs each with more than 100 young within; stingrays rose from the sea floor past black grape-sized bunches of cuttlefish eggs and the mermaid's purses or egg cases of the smooth-hound sharks and dogfish that abound in these forests. And the reason for all this life was the kelp.

Sussex's kelp forest contained four main species. All grow quickly, reaching a height of between 1 and 4 metres. The oarweed, with its short stem, or stipe, branching into blades with between five and twelve 'digits', and looking a little like a stubby hand with long waving fingers, tends to grow in the shallower water. Moving slightly further offshore you could find sugar kelp, tangle and furbelows. The tangle kelp looks similar to oarweed but has a much longer, more ridged stipe. But sugar kelp and furbelows are very different. Sugar kelp is a single long frilly frond draping the sea floor. The delightfully named furbelows has a bulbous holdfast that looks a little like a strange swollen yellow-brown fruit on the seabed, with a long, flat stipe and a mass of kelp fingers reaching out as if trying to gather the rays of sunlight that all kelp need if they are to grow. Together, they formed a forest said to have extended at least a mile out to sea and so thick that, in the shallower parts, people in rowing boats struggled to push their oars through it, and even propeller-driven boats got stuck.

Regular seasonal storms are vital contributors to a kelp forest's life cycle. It is then that many species of kelp release their spores in the winter to allow the disturbed waters to disperse them. But when a particularly big storm comes through it can break the stipes and holdfasts of some species of the longer-lived kelp as well as gathering the fronds of other species that naturally die back, creating piles of kelp all along the Sussex coastline. These kelp mounds divided local opinion. Some farmers collected the kelp to use as a free rich source of fertiliser. Nature lovers delighted in the rich offshore environment it represented, but many people felt that kelp was fine to have in the water but not exactly appealing when washed up on the shores of towns that hoped to attract tourists to their beaches. Questions were even raised in Parliament, Labour peer Lord Aylestone, who lived on the south coast, enquiring of Her Majesty's Government 'what help they are giving to local authorities, in the interests of tourism, to deal with decaying seaweed on the beaches of holiday resorts such as Worthing, West Sussex'. But in general there was a balance. The kelp forest was part of life along the Sussex coast; it always had been and it always would be. Until 1987.

The storm was so severe that the mounds of kelp on the beach were bigger than anyone could remember. But it wasn't this that destroyed the forest in the long term – it just made it easier for those that did. A different kind of storm had been brewing.

Although there had been trawling in the region since the 1850s, new, more powerful trawling boats, whose owners were emboldened by changes to fishing legislation, moved in. During the exact period when the kelp forest needed time and space to recover from the storm, heavy-weighted nets and chains were dragged along almost all of the coastline. Mussel beds were smashed, most of the remaining kelp was destroyed and much of the life that once lived there was either caught or dispersed. Time and time again the trawlers passed through churning up the bottom, and by the time a detailed survey was undertaken nearly all the kelp was gone from a vast area of coastal waters in what was supposedly one of the most conservation-minded countries in the world.

It is unlikely that trawling alone destroyed the kelp forest – industrial pollution, sewage and sediment dumping almost certainly played a big role too. But by trawling much of the seabed, year in year out, recovery was impossible.

Years passed. After a couple of decades memories had faded and it was easy to take a walk along Worthing beach or swim at Selsey and not realise much was wrong. It was still nice to escape the hustle and bustle of everyday life and be down by the sea. Over time one can forget the flocks of seabirds that used to soar overhead or dive for fish, the variety of shells washed up on the beach, or how regularly you saw dolphins leaping out of the water in the distance while you were on a morning stroll. It is well documented that over time we all become used to a nature-depleted world and forget the riches we once had. Marine scientist Daniel Pauly called it shifting baseline syndrome. It has been described as 'environmental, gener-ational amnesia', but Pauly spoke of it in more forgiving and accurate terms, as 'a gradual change in the accepted norms for the condition of the natural environment due to a lack of experience, memory and/or knowledge of its past condition'. In other words, what we see today we tend to think of as normal, as the 'baseline' by which we perceive improvement or decline. And it leads us to drastically underestimate what a healthy ecosystem should be. What we often think of as 'healthy' or at least 'normal' today would look much diminished to a previous generation. We get used to a degraded environment and, as a result, too often we don't even aspire to recover what we once had.

But then, in 2019, three very different lives began to overlap.

Dr Sean Ashworth grew up in Brighton, just a few miles from Shoreham. The 1987 storm had blown the roof off his bedroom. Fortunately he was away at the time but he remembers returning a couple of days later to a changed landscape – little did he know that far greater destruction was just beginning offshore. He had known these coasts all his life and now he was working at the Sussex Inshore Fisheries and Conservation Authority (IFCA) hoping to care for them. He repeatedly heard tales from fishers of their dwindling catches and struggles to make a living, and this chimed with the fish catch data the IFCA was gathering. Fond of the line 'you can't have fisheries unless you have fish', Sean was searching for a way for sustainable fisheries to thrive in his home waters.

Eric Smith, a freediver now in his mid-seventies, had been snorkelling and diving in these waters since he was eleven years old. While knee and hip replacements mean he doesn't walk as well as he once did, Eric has the youthful energy, bright eyes and unruly hair of a man who has spent a well-lived life in and on the water. Eric had kept records of every dive and could look back and chart the transformation of his haven into horror. Heartbroken at the sudden destruction of the marine world he loved, Eric realised he was one of the few people who had actually seen this with his own eyes. He personally knew of mussel beds that had vanished, and had swum through kelp forests where now there was a desert of sand and sediment. For years he had seen significant numbers of commercially valuable species such as lobster, sea bream and bass on almost every dive; now he would be lucky to see a bucketful in a year. Two of his diary entries from April 2005 bring home the speed of the transformation to his beloved coast.

17 April

We found the bream quite quickly, spotting their moon beds dug into the chalk from the surface. The sea was alive with fish, the smaller bream were doing the digging, fanning the bottom to create the hollows and the bigger fish moving in to lay their eggs or fight over the right to. Undulated rays were lying around the outskirts of the beds feeding on the worms and other small crustaceans disturbed by the bream and the vast shoals of pollack up to two kilos drifting above the scene. We lay looking at this panorama with cuckoo wrasse nosing around us, with bass coming through in shoals, some weighing in at eight pounds. We left feeling good at seeing the sea so alive with fish.

24 April
Returning to the same site
The sea was calm, and we had started at first light under the old adage that the early diver catches the fish. As we pulled the boat up onto the promenade two pairs of

trawlers could be seen towing out to sea at an angle alongside the main reef less than a mile out. Slipping into the water 30 mins later a scene of total devastation could be seen below us in the crystal-clear water, no fish could be seen in our field of vision, the bream beds so prolific the week before were wrecked, the edges of the beautifully dug pits had been dragged down and large boulders had been dumped into the middle of them. Torn weed lay about the bottom and all that remained were a few very small bream trying to rebuild their nests. We followed the trawlers that were now some two miles ahead finding hundreds of small bass and bream floating dead on the surface, having popped out of the now full nets.

Others may understand from scientific studies the riches this coastline once offered; Eric knew what it actually felt like to swim through them.

And so did Sarah Cunliffe. Now an award-winning wild-life filmmaker specialising in underwater films, frequently set in far-flung locations, Sarah was most at home on the grassy downs and stony shorelines of the Sussex coast. Most days she walked her long-haired Jack Russell named Spot (after the distinctive brown patch on her rear) on these beaches, and in the misty early mornings she had looked out towards the kelp forests on the harbours off Selsey where she had conducted research on the sea life for her final-year degree project. These extraordinary Second World War structures were originally planned as floating harbours with a crucial role in the D-Day landings – Selsey was one of two sites where they were secretly built by the Allies under the codename Mulberry. Today the remains of some of these constructions, which were broken up or never used, have become artificial reefs that are popular for snorkelling and scuba diving due to the frequent aggregation of fish that still occurs around them – shoals of bib, hundreds strong, are regularly seen, as are species such as pollock and bass. However, despite these attractions the area is today without the thick kelp forests that were the constant backdrop during the time of Sarah's research.

Eric had spent decades documenting the damage caused by trawlers and writing letters to people he thought could help end it. Sarah had spent decades building a thriving business filming the ocean all around the world. As such she was well aware of the precarious state of global marine life, but like most people in southern England (including this book's two authors) she was unaware of the destruction of her own patch. Sean had spent the last decade monitoring fish populations and trying to work out how to use the law to generate a sustainable fishing industry thriving in healthy seas. Together they held the first jigsaw pieces of what would become a beautiful picture of hope.

Eric shared with Sean his personal testimony of the vast flourishing forest of the recent past and the destruction he had seen. The nature lover in Sean longed to see such sights for himself; meanwhile the pragmatist in him perked up at the sound of the commercially valuable, fast-growing fish species that Eric could testify had once abounded here, and the ability to create sustainable fisheries. Sean worked with the team at Sussex IFCA to put the science together with the anecdotal evidence and create a highly ambitious plan for a by-law banning inshore trawling along almost the entire East Sussex coastline. Their plan didn't just involve stopping trawling in sight of shore but as far as 4 kilometres out to sea all the way from Selsey to Shoreham, with a further, narrower zone from Shoreham to Rye – 300 square kilometres of ocean completely free from trawling.

It is worth pausing for context to illustrate just how revolutionary the proposal was. At this time only 1 per cent of the entire waters of the UK were properly protected by law. Even though on paper almost a quarter of the seas were labelled marine protected areas, nearly 97 per cent of those protected areas were still bottom-trawled. Pushing for a by-law to create a no-trawl zone of this size was, therefore, not only visionary but also very ambitious.

This was exactly Sarah's thought when she first met Sean in the summer of 2019. She had been to see him to discuss a completely different matter and was shocked to discover from Sean, almost in passing, that the kelp forest she had studied as an undergraduate in the early 1980s had gone. A patch of ocean Sarah once looked at most weeks of the year had had the life ripped out of it without her or anyone else she knew even realising. Her first thought was to fight back; her second was 'no one knows or cares about kelp'.

Sarah set out to change that by starting a campaign called Help Our Kelp. She brought on board established conservation groups like the Sussex Wildlife Trust and Blue Marine Foundation, launched a PR and social media push, and started getting attention for a forest that no longer existed. A film she made called *Help Our Kelp* spread fast around the region and beyond. It helped the campaign garner unparalleled attention and interest for a local public consultation on a by-law. Sussex Wildlife Trust took on a key role, chairing Help Our Kelp and bringing their membership and expertise to bear.

In parallel, the organisation Sean worked for, the IFCA, held a wide consultation and submitted detailed proposals to the British government. One of Sean's most significant pieces of evidence had been retrieved from a skip! Several years earlier he had rescued a handful of reports that were considered obsolete and were being thrown away. One of them was a local council study from the 1980s mapping the then flourishing kelp forest off the West Sussex coast. Ironically, it was commissioned at the time because the council viewed the kelp as a nuisance and wanted to know how big a problem it was dealing with; now it was a vital document showing clearly and scientifically what had been lost.

Sean's evidence, coupled with the attention generated by Sarah and the Help Our Kelp campaign, led to unprecedented public support for the proposals, and the burgeoning movement, forged from frustration, began to be fuelled by hope. Little by little the communities along the coastline started to hear about the lost treasures they could perhaps regain. But major proposals like this one are not going to be uniformly welcomed. There were some who wanted trawling to continue, in many cases quite understandably as they believed their business depended on it – though Sean notes that the most vociferous opposition came from big trawling businesses based far away.

Nonetheless there were enough in the fishing community who could see the writing on the wall. Fish stocks were too low. More and more effort was going into catching less and less. From scientific research around the world it is now widely understood both that kelp forests function as critical marine nursery grounds and that the density with which kelp grows buffers the energy of the waves so that egg cases which are laid on the kelp are not dislodged and washed away. Not only could there be no fisheries without fish; there would be few fish without kelp.

In March 2021 the UK Secretary of State for Environment, Food and Rural Affairs, George Eustice, approved the new by-law and the no-trawl zone could go ahead. Speaking immediately afterwards, Sean described it as 'a landmark decision for the sustainable fisheries of Sussex, a landmark decision for the marine environment . . . and it could be a landmark decision for the marine ecosystem around the entire country'. 'Landmark' is a word sometimes used too lightly, but on this occasion Sean cannot have dreamed just how right he would be.

While life in the ocean can often recover far faster than on the land, it still takes time. And if this project was going to be a true landmark then the science evaluating its effectiveness needed to be robust. But ocean recovery is not easy to calculate. There is no one technique or measure that can capture everything that is happening beneath the waves. An array of researchers and scientists deploy a variety of tools, from baited remote underwater video cameras that assess which species are using which areas, to cameras towed underwater on fixed transects at regular time intervals that record what is growing or living on the seabed. Other experts use DNA analysis to discover which species have passed through an area in the preceding three weeks. They collect samples of the ocean and push it through a DNA syringe that separates the seawater from tiny traces of DNA that animals leave behind in the water, such as scales, faeces or skin cells. In this way you get a genetic fingerprint of species that you might never capture on camera.

Researchers from local universities who were interested in this unique opportunity to monitor ocean recovery from its very beginning expressed a desire to conduct baseline studies if the trawling ban was approved. The rest of the jigsaw pieces were coming together. Crucially for both the kelp forest in the future and for persuasion in the present, researchers surveying a location called Pullar Bank, just off Selsey, discovered a large stand of kelp in an area that had never been trawled because it was too shallow. This stand was of mature kelp, so each winter they would release thousands of spores which could one day aid in regrowing the forest if the trawling ban succeeded and the water conditions were good. But maybe more importantly it was also an opportunity for Sarah to film this remaining old forest and show people what the future could look like for the whole of their coastline.

In the early stages of recovery across an area too wide for scientists to monitor constantly themselves, citizen science plays an important role. Divers are encouraged to record what they see on each dive. Beachgoers are requested to send photos of interesting sightings such as washed-up kelp, shells or eggs which might indicate changes in the nearby seabed. By utilising a wide range of methods such as these, scientists can tell which species have arrived, and the order of their arrival, over how long a time frame and in what numbers. It is invaluable data for the growing science of marine restoration.

It is early days still, but there are encouraging signs that the sea off Sussex is beginning to heal. Eric has seen huge new mussel beds forming on the previously dredged sea floor. These bring a twin benefit. First, the mussels themselves clean the water – each individual mussel filters approximately 25 litres a day, so large mussel beds provide cleaner water, which lets through more sunlight, which, in turn, helps kelp to grow. Secondly, the way that mussel beds grow – with hundreds of thousands of mussels densely multiplying, spreading outwards and even growing on top of old mussels – creates a hard base to which kelp plants can attach their holdfasts. In the absence of an already rocky sea floor, expanding mussel beds can be thought of as creating the ground on which the kelp forests can grow. The first mussel beds were the size of tennis courts, then football fields, and now we have run out of comparable sports pitches.

To keep the community aware of progress Eric, his daughter Catrine and many other local freedivers formed a group to share images and arrange school talks. They left weighted cameras on the seabed and have recorded black sea bream nesting in historic sites where they had not been seen for years. Smooth hound sharks, blenny and many other species, including an exceedingly rare angel shark, john dories and electric rays, have been recorded. Paddleboarders have sighted undulated rays over 400 metres offshore, lobsters appear to be returning to previously barren sites and dolphins are being sighted most days – a strong indication of improved conditions. Scientists report high numbers of fish fry such as herring, sardines and sand eels in the area, and there have been sightings of kelp washed up on the shoreline for the first time in years.

Some fishers who had given up their trade due to falling catch have started to return to these seas using low-impact fishing methods, and Sean is hopeful that a sustainable fishing industry can grow alongside the kelp forest. All parties are clear that fishing is an important part of this recovery. Small-scale nets and pots operated by the local fishing community will bring jobs and add to local pride in the recovery of the kelp forest.

Sarah, Eric and others closely involved are quick to acknowledge that it is too soon to expect conclusive scientific proof that the forest is returning. But they are equally quick to point out that everyone who knows these waters has already noticed changes in just a couple of years. They raise concerns that intensive but legal sediment dumping just offshore might hamper recovery, as it reduces the light reaching the kelp, but public opposition to this practice is rising faster than the sediment is falling. Help Our Kelp has become the Sussex Kelp Recovery Project, showing a gear change from an urgent campaign to a long-term programme of restoration. It seems that the coastal communities of Sussex have taken their ghost forest to heart, and they want it resurrected.

When the trawling ban went from idea to law, Sean voiced a hope that it could be 'a landmark decision for the marine ecosystem around the entire country'. The first signs are that he might just be right. Other councils across the UK have seen the cooperation in Sussex between local government agencies and the community, witnessed the pride and excitement it has generated, and are looking to follow suit. Could it be that local action succeeds where decades of national policy has failed?

Surprisingly, it has even had an impact many miles inland. Farmer James Baird saw what was happening and realised that while land and ocean are very different realms, what happens on one may have an impact on the other. He has started an initiative called Weald to Waves to create a recovery corridor of over 20,000 hectares of contiguous habitat. When completed this will link farmland, bird reserves, forest and rivers with the newly protected seas of East Sussex.

It is just the beginning for the Sussex kelp forest. But it is also just the beginning of a new attitude in the UK where local communities are acting to protect and restore their own seas rather than waiting for instruction from the government. Finally, the proud maritime nations of the UK might go beyond talking about marine reserves and long-term sustainable fisheries, beyond protected areas that only exist on paper and into a new era of a thriving wild ocean that can support vibrant coastal communities.

5

ARCTIC

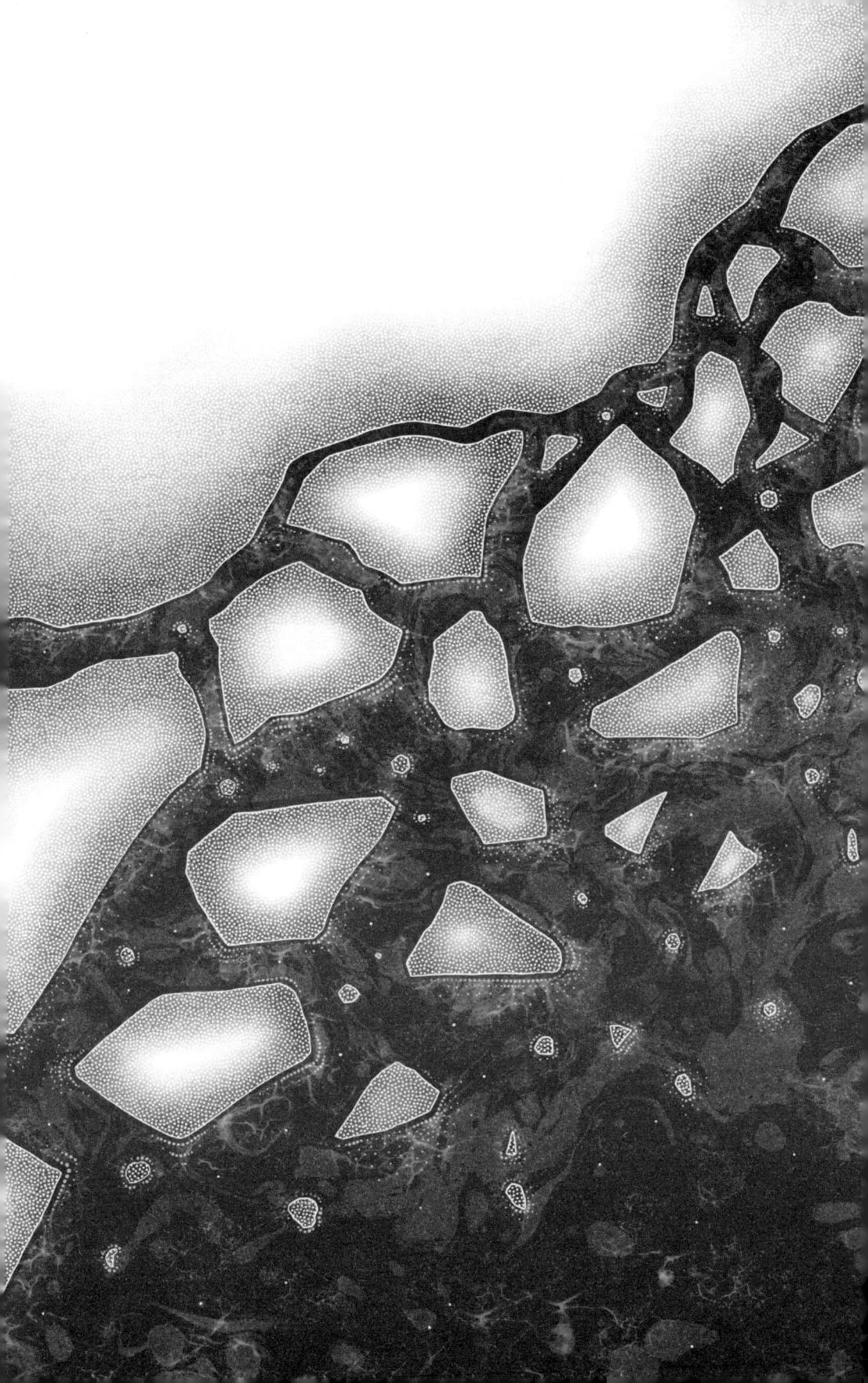

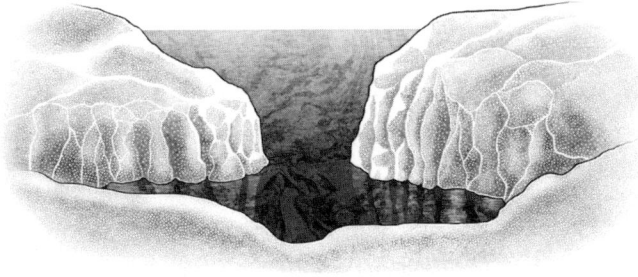

LIFE UNDER THE ICE

I t is strange now to look back at a long life spent recording the wild places of our world. Often I reflect on my immense good fortune, or laugh at the memory of great adventures with good friends. Such a long career also means that I have discovered whether or not many environmental hopes and fears have actually come to pass. The recovery of the great whales or the regrowth of the forests of Costa Rica would have surprised and delighted my younger self. Many other places and species that we all thought would disappear under humanity's relentless pressure have survived despite the odds – often, it must be said, thanks largely to the tireless efforts of members of the same species which threatened them in the first place. These are heartening things to reflect on and give me hope in these challenging times for our natural world.

But only the Arctic gives me both a longing to experience its other-worldliness one more time and the knowledge that this is impossible, as it is no longer the place I once visited – nor will it be again.

'The poles are permanently capped with ice,' I say in 1983's *The Living Planet*.

Much later, while standing at the North Pole for the original *Frozen Planet* series in 2009, I am filmed speculating that the North Pole could become occasionally ice-free within the next few decades *if* we do not slow global warming (as we called it then). Today, some fifteen years later, it is clear that the sea ice of the Arctic, including that over the North Pole, will disappear completely during the summer months and no amount of human effort can arrest that. We have passed that point. Of course we must still do all we can to avert even more extreme warming, but for the frozen Arctic Ocean that I once stood on, it is now a case of *when* not if.

It is understandable that much of our attention has focused on the people and wildlife impacted as the ice melts, but filming techniques I have had the privilege to watch develop over my career have also revealed the beauty and importance of the ice itself for life in the Arctic.

The North Pole has only one sunrise a year. At the March equinox it rises low in the sky and in the September equinox it 'sinks' below the horizon and darkness returns. Photosynthesis is impossible without sunlight so in winter algae cannot grow. But in spring and summer the 24-hour daylight results in rapid growth. Surprisingly, this process actually starts in fresh water. Although the sea ice is a frozen ocean it is largely made of fresh water, as during the process of freezing the salt is pushed out of the seawater. So when the melt begins in the spring it forms pools of fresh water on top of the ice which can be full of nutrients and excellent places for the growing algae. The sea ice itself provides a base for growing algae, much as soil does for plants on land. Zooplankton eat the algae, and so form the foundation of many food chains in the Arctic. As a simple example: Arctic cod eat zooplankton, seals eat the Arctic cod and polar bears in turn hunt the seals.

During my years of filmmaking ever more stunning images have been captured by highly skilled divers portraying the world under the ice. The colours are quite extraordinary. The Arctic sea ice isn't a uniform sheet but has channels, gaps and ridges, so some parts let through far more sunlight than others. This creates wonderfully varied light but also governs where algae grow and in what volume. So in a single shot looking up from under the ice, you can see deep red undulating ice formations penetrated by bright clear shafts of light. It is quite breathtaking.

Filming underwater, and in particular under the ice, has always been tricky, but technological innovations have enabled us to record footage that would have been unthinkable just a few decades ago. On more recent series expert divers use rebreathers – closed-circuit systems which recycle a significant part of the diver's breath by removing the carbon dioxide. As well as extending the amount of time the diver can stay underwater, it also means there are no bubbles to disrupt the shot or noise to disturb the wildlife. And that means we can seek out some truly extraordinary behaviour.

For *Planet Earth III* it took months of daily dives beneath the ice to reveal that a hunt between two small pteropods (swimming sea snails and slugs) can be every bit as exciting as watching polar bears or orcas.

A sea angel is a type of sea snail, so named because of the wing-like structures with which it propels itself through the water. These 'wings' are an evolutionary adaptation of a land snail's 'foot'. The snail is only 1 to 2 centimetres in length but nonetheless is a formidable predator, and its favoured prey is another form of swimming snail – the sea butterfly. Sea butterflies are relatively abundant in the Arctic Ocean, feeding on the phytoplankton and zoo-plankton. Neither the sea angel nor the sea butterfly can see but they are adept at sensing movements in the water. The sea angel has the extraordinary ability to turn its mouthparts inside out to form a trap of tentacles with which it can ensnare a sea butterfly. It is a fascinating sight to witness, and it has only been possible for us to do so because of the life-rich waters of the Arctic and the dedication and talent of scientists and camera operators.

While the technology has evolved, filming under the ice has always required great bravery and skill from underwater cinematographers at the top of their game. I confess I have never much fancied slipping through a seal hole, having your regulator freeze to your lips and navigating under the ice, knowing that if you lose your way you may never find a route to the surface again. But that makes me admire all the more those who did so for us.

The images they have recorded have revealed that ice is not merely a platform on which polar bears can hunt; it is the foundation of life in the Arctic. We may never again witness the summer sea ice I once knew but there is still the opportunity to ensure that Earth retains winter Arctic sea ice and frozen land at the north together with the rich abundance it provides locally and the stability it gives to our entire planet.

DA

In 2007 in the remote Alaskan Arctic an Inuit hunting party discovered a strange-looking spear lodged in the flank of a bowhead whale. It was clearly very old, but where had it come from? How long had the creature carried it around the Arctic Ocean? They sent it for analysis to experts who astonishingly revealed that the spear had been thrust into the whale towards the end of the reign of Queen Victoria, in around 1890. It appears whalers from New Bedford, Massachusetts, had attempted to kill the whale, but the blow was not fatal. The bowhead then spent the next 115 years going about its life in this most northerly of ocean basins with the top of a spear in its body. Given that whalers would have been unlikely to waste effort hunting a juvenile, the bowhead was probably already an adult when attacked, making it over 120 years old, though there is reason to believe bowhead whales can live even longer than that, perhaps up to 200 years.

The bowhead is a true Arctic species, and is typically

found in waters that experience at least some seasonal ice coverage. Under its dark skin is blubber that can, remarkably, be almost 50 centimetres thick – the thickest of any whale. It is named after its extremely large head, which can be a third of its entire body length, and it uses its bow-shaped skull to break through sea ice, allowing it to search for food in areas where other marine mammals cannot go due to the risk of finding themselves too far from a breathing hole in the ice. The distinctive white chin is located below the largest mouth of any animal, filled with dark baleen plates up to 4 metres long through which they filter krill and other invertebrates. Sound is important for communication, mating and locating food, and bowheads are famed for their far-reaching 'songs', which can travel for many kilometres through the cold water. In short, the bowhead is wonderfully adapted for a life in the Arctic Ocean.

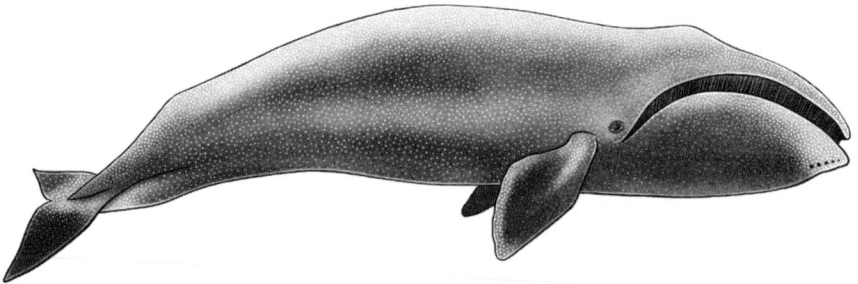

The bowhead whale, showing its characteristic mouth.

While the spear found in this particular whale was a one-off curiosity, the body of any reasonably old bowhead would reveal the great changes that have altered the Arctic Ocean over the last 100 years. Scars from ship strikes and entanglement in fishing gear are useful to help scientists identify an individual whale, but they also hint at a more traumatic existence than the whale's graceful gliding through the cold water suggests. The fatty parts of their body may contain persistent organic pollutants – dangerous, hard to break down chemical compounds – brought up to the Arctic Ocean on air and water currents from poorly disposed waste in Europe and America. The race is now under way to understand their home as it changes fast around them, and to take what action we still can to protect it.

It is tempting to lump our planet's two polar oceans together and assume similarities, but they are, in fact, profoundly different. The Southern Ocean is a giant ring of sea circling the Antarctic land mass whereas the Arctic Ocean is centred around the North Pole and surrounded by land. This major difference defines their wind and wave conditions, the currents they experience, how much water they exchange with the rest of the world's ocean and even how salty they are.

The Arctic Ocean is the smallest of the world's five ocean divisions and is located over a deep depression known as the Arctic Basin. It is usually defined as the part of the world's ocean that is within the Arctic Circle – an area marked by the northernmost point where the sun remains visible on the winter solstice, around 21 December, and the southernmost point at which the midnight sun can be seen on its summer solstice six months later. Encircled by land owned by five Arctic coastal states – Canada, the kingdom of Denmark, Norway, the Russian Federation and the United States of America – it forms part of the home of over 4 million people including approximately 500,000 from forty different indigenous peoples. Its near enclosure by land, and its resulting few points of connection with the rest of the world's ocean, is an important factor in shaping the life that lives here.

Viewed from above, it is clear to see that there are just two links to the outside ocean: the connection with the North Atlantic, and the tiny opening of the Bering Strait between North America and Russia connecting the Arctic to the Pacific Ocean. Looking underwater reveals that this second opening is even smaller than it initially appears, for the Bering Strait is extremely shallow – typically less than 100 metres deep. Indeed, until quite recently, in geological terms, there was no gap at all. During the height of the Pleistocene, some 20,000 years ago, so much water was locked away in glaciers and massive ice sheets that the sea level was well over 100 metres lower than it is today and the Bering Strait was the Bering Land Bridge – something of a misnomer perhaps, as the 'bridge' would have been a vast grassland steppe up to 100 kilometres wide supporting the migration of plants, animals and people between America and Asia; nonetheless the Arctic Ocean would have been completely closed to the Pacific. Today, though, this small gap allows for the exchange of waters, ships and wildlife.

More significant in both size and impact, however, is the place where the warm water of the North Atlantic Current sweeps up from the Caribbean, through the Sargasso Sea, between western Europe and Greenland and hits the Arctic archipelago of Svalbard. This is the major point of connection between the rest of the world's ocean and what has become known as the upside-down ocean. The Arctic Ocean was given this nickname by the explorer Fridtjof Nansen, who deliberately froze his ship in the Arctic sea ice knowing that the movement of the ice drift would carry him and his crew from Siberia back to their destination in Norway. During this time he took water samples and, in discovering cold water on the surface and warmer saltier water below, declared it to be 'upside down'. Nansen knew from previous studies that, in most parts of the ocean, evaporation removes fresh water from the surface, leaving warmer, saltier water on top and cooler water below. But not so in the Arctic. The reason for its upside-down structure is the huge volumes of fresh water coming into the Arctic Ocean from melting sea ice and rivers, making the Arctic the freshest of all ocean basins. Fresh water is lighter than salt water so stays near the surface.

In fact the Arctic Ocean can be thought of as consisting of three distinct layers. Just below the freshwater layer is a vital barrier known as the Arctic halocline. This layer of cold salt water lies between the cold, fresher surface water and the warmer, salty deep layer. The halocline exists because when sea ice forms it releases the salt, which cannot freeze. This salt makes the surrounding water saltier and denser, causing it to sink below the fresh water at the surface and acting a little like a lid on the warmer deep water below. If it did not exist, the surface of the Arctic ocean would be warmer and the Arctic sea ice would be melting even faster than it is today.

The incredibly rich wildlife of the Arctic Ocean is not evenly distributed, and it is in part this unusual mix of water temperatures and salinity that is responsible for this. Where the warmer waters of the Atlantic and Pacific flow into the Arctic Ocean through relatively narrow entrances, they drive a mixing of waters that brings vast quantities of nutrients towards the surface. These nutrients feed plankton at the base of the food chain, to the benefit of the rest of the wildlife of the Arctic Ocean. Great numbers of fish, seabirds and marine mammals congregate to feast in these regions, and that is one of the reasons why much of the most spectacular wildlife footage you will see from the Arctic Ocean is either along the Atlantic arc stretching from Baffin Bay through Svalbard to the Kara Sea, or near the Bering Strait. A second reason is that these areas also have large rivers flowing into them bringing more and different nutrients. (Admittedly a third reason is that these areas are simply more practical to film in!)

Further towards the centre of the Arctic Ocean, over the Arctic Basin, marine life is less abundant. This is partly due to being further from the highly productive waters at the land gaps where the Atlantic and Pacific meet the Arctic, but also because this area has more persistent sea ice. The ice blocks the spring and summer sun from the water, limiting the growth of phytoplankton which needs sunlight to photosynthesise and grow. Without this growth there is much less biomass at the bottom of the food chain and so fewer animals higher up. However, knowing this also gives us clues as to where we can find wildlife in these parts.

A single 2-metre-long tusk breaks the surface, followed by a dark head and white speckled chin. A narwhal, perhaps returning from a dive hundreds of metres down where it was hunting for cod, squid or halibut. It is far from land, surrounded by ice. It can hold its breath for over twenty-five minutes, but being a mammal it still needs to surface. And so this narwhal has found a 'lead' – a large fracture in the sea ice it uses for navigation and, of course, in order to breathe.

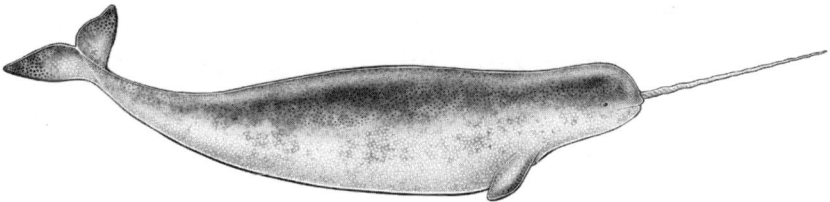

Only male narwhal have a tusk,
an enlarged canine tooth attached
to the left upper side of the jaw.

While narwhal can be found in large herds near shore in summer, they move far offshore when the freeze begins. Out here they are safer as the ice provides excellent protection from their main predator: orcas. Narwhal have no dorsal fin and so can hide right up under the ice; orcas, on the other hand, have a very large dorsal fin, which is more awkward in this environment. It has been this way for millennia but now, as the sea ice diminishes, the orcas' range is increasing and narwhal have fewer areas to hide.

Leads and other gaps in the ice such as polynyas – areas of open water amid the ice created by warm water upwellings or persistent winds – also provide refuges for a variety of Arctic wildlife. Here, the first sunlight in spring can reach open water and trigger plankton blooms, kick-starting growth throughout the food web. Some polynyas are one-offs created by unique local conditions in a particular year, but there are others that remain open year round and even some which occur at the same time and same place each year, forming a reliable destination for the wildlife of the Arctic Ocean.

This is a world where narwhal traverse gaps in millions of square kilometres of sea ice, where centuries-old whales call across vast distances; a place of ancient land bridges and long months of darkness. It can feel far away from human cities and centres of industry. Yet even here our industrial world is pervasive, creeping, ever present. In the 1950s pilots flying over the North American Arctic noticed a strange haze on the horizon. No one knew what it was or why it only appeared at certain times of year. Decades later the haze was identified as aerosols of sulphate and soot originating from European heavy industry. In the boreal winter, large parts of Eurasia are in the path of winds that flow up to the Arctic, so airborne pollutants from factories, cities and agriculture are transported north and approximately two weeks later they reach the Arctic.

Five large rivers – the Mackenzie and Yukon in North America and the Lena, Ob and Yenisey in Asia – each with immense catchment areas and countless tributaries, flow into the Arctic Ocean. Numerous small and midsized rivers also end their journey here. Like the winds, they carry life-giving nutrients as well as contaminants from settlements, mining, agriculture and industry which wash out into the Arctic Ocean, often in pulses when the ice melts and the rivers surge. Meanwhile the Gulf Stream flows into the North Atlantic Current, which then sweeps north past Europe, collecting on its way water-soluble contaminants such as compounds of heavy metals, plastics and pesticides, delivering them to the Arctic Ocean. In these ways chemical substances that have never been used in the Arctic arrive there.

Persistent organic pollutants (POPs) is the term given to poisonous chemical substances that take a very long time to break down and therefore endure for decades or even centuries in the environment. As a result they can bioaccumulate, that is, build up in the fatty tissues of animals. They include certain pesticides, insecticides, flame retardants and solvents. The most infamous is DDT, an insecticide banned in many countries in the 1970s but still detected today in the blood of children born decades after the ban. The risk from POPs appears to be greatest for animals living towards the top of the food chain. Populations of orcas, polar bears, walruses and ringed seals have all been tested and found to have high levels of POPs in their fatty tissue, muscle and organs. While the pollutants may only be present in very small quantities in the fish a seal preys on, because it eats many fish these pollutants gradually accumulate within its body. A polar bear or orca eating many seals a season will then inherit all of their toxic legacy. These pollutants are linked to cancer, disruption of the hormone and reproductive systems and damage to the nervous system in both humans and animals. Thankfully many have now been banned and regulation is improving in much of the world, so although there is still further progress to be made we can at least now envisage a time in the future when we no longer need to ask questions such as: why does a polar bear contain flame retardants?

But our most complex and pervasive impact in the Arctic comes from what initially appeared to be a far more benign chemical compound: carbon dioxide. In summer 2017, in the far Russian Arctic, 100,000 walruses hauled out on a remote beach. At this time of year they would usually be resting on the sea ice above the continental shelf. They would flop off the edge of the ice and descend to a depth of up to 80 metres to forage on the shallow sea floor, using their highly sensitive whiskers to locate shellfish and crustaceans. Walruses have developed ingenious tactics for reaching hard-to-access food. They can shoot jets of water from their mouths to dislodge prey and can create such strong suction that they are able to suck shellfish such as clams right out of their shells. Once sated, they use their tusks to help pull their 3-metre-long bodies, weighing approximately 1,500 kilograms, up on to the ice to rest, earning them the delightful scientific name *Odobenus rosmarus*, or 'tooth-walking sea horse'. (Their common name – walrus – is thought to come from old Dutch sailors and translates as 'shore giant'. For once, the scientific terminology is the more amusing!) The ice needs to be reasonably thick to support thousands of walruses, which then lie on it practising an astonishing range of vocal acrobatics – from whistles and honks to rasps and bangs, depending on the intention of the singer.

But the ice is no longer where it once was. It has retreated further north in summer and now is more likely to be over deeper water than the continental shelf where walruses mostly feed. Unlike their seal relatives, walruses cannot stay at sea for long periods of time. They must rest. Without ice they cram on to beaches, but though agile and efficient in the water, they are cumbersome and poor sighted on land. In 2017 there were so many on this beach that hundreds were forced on to the cliffs to find space. Walruses are not natural climbers but somehow they made their way up. Their eyesight is poor out of water and upon hearing some of their herd in the water below, perhaps preparing to leave, they headed in the most direct route towards them. Hundreds fell to their deaths. Their world is changing so fast that their senses and habits are simply unable to keep pace.

Thankfully these beach haul-outs do not often end quite so horrifically, but nonetheless they are having an impact in various ways. Calves, in particular, are often killed when the herd gets startled by a polar bear or a passing boat and then stampedes. Then there is some early research indicating nutritional problems, almost certainly because a beach concentrates many animals in one location, depleting food sources and making the walruses swim further to find food. Drifting pack ice, meanwhile, naturally keeps them moving to new feeding grounds as well as providing them with a far wider range of haul-out locations.

The Arctic is warming more quickly than any other region on the planet. Studies show it to be heating between two and four times as fast as the global average. This is known as Arctic amplification and is due to a range of factors, one of the most significant being albedo. Albedo is the amount of light hitting a surface – in this case ice – that is reflected back into space, thereby keeping the temperature in the region cool. As the Arctic heats, there is less sea ice; this uncovers darker water underneath with a far lower albedo, so there is less of the sun's energy reflected back into space while the water absorbs the additional heat. The Arctic Ocean therefore heats even more, with even less ice to reflect it, causing more ice to melt. And so the vicious cycle continues.

We are only just beginning to see what these changes might mean for the wildlife and people who rely on the Arctic Ocean. Orcas appear to be benefiting, at least for now. Historic records suggest they didn't venture into parts of the Arctic Ocean such as north-eastern Canada because of the sea ice, but as the ice melts so they are expanding their range into completely new areas and taking advantage of better prey opportunities. Bowhead whale

A sea angel is one of the many creatures that abound in the cold, rich waters of the Arctic Ocean. Only 1 to 2 centimetres long, it is nonetheless a formidable predator.

A pod of narwhal gathers at a break in the sea ice in Nunavut in the Canadian Arctic. The region has recently acquired the name Last Ice Area since it is forecast to be the final place in

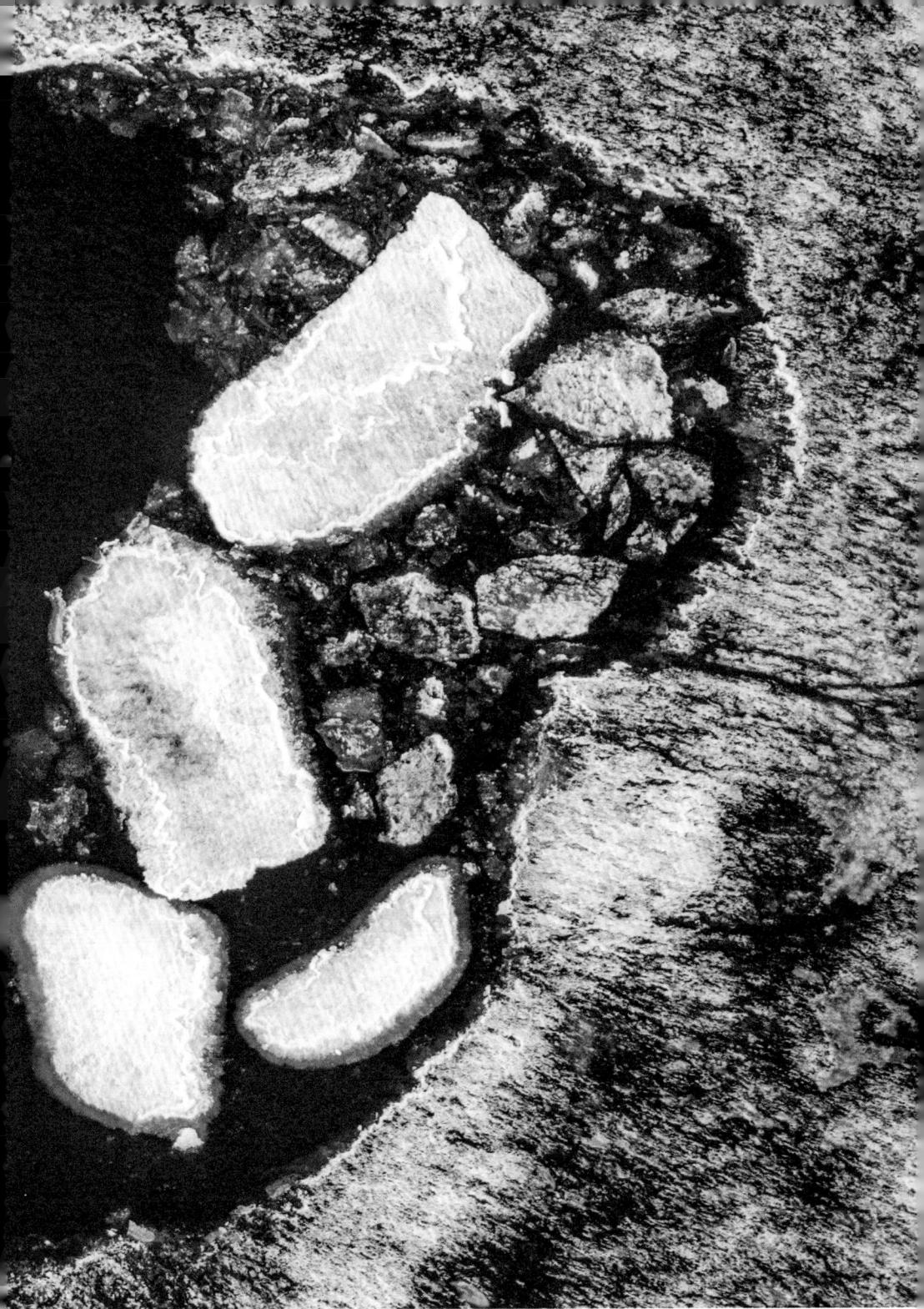

the Arctic to lose its summer sea ice as our planet heats.
Nunavut is also home to two massive, pioneering marine
protected areas under the stewardship of the Inuit population.

Walruses just off the Arctic island of Spitzbergen. They typically use their strong tusks to help haul themselves out onto sea ice to rest between feeds, but are increasingly found on beaches, often in vast numbers, as the availability of sea ice thick enough to support their 1,500-kilogram bodies diminishes.

Orcas are among the most intelligent and versatile hunters in the ocean. As Arctic sea ice declines, they are now able to enter new regions, hunt different prey and compete with other predators.

calves, narwhal and belugas, on the other hand, are already finding that less sea ice means less protection from orcas. Populations of marine mammals who have rarely, if ever, encountered orcas before, and therefore have no learned evasion or defence behaviour, will suddenly find themselves being hunted by perhaps the most accomplished and intelligent predator in the ocean. Indeed, it is thought that over time orcas may replace polar bears as the dominant marine mammal predator in parts of the Arctic.

If the heating stopped now, perhaps species could adapt and a new equilibrium could settle. A fascinating study in 2018 gave a window into a different Arctic Ocean. It compared the diets of belugas, ringed seals, Greenland halibut and Arctic char in 1990–2002 and 2005–2012. All significantly changed what they ate as the Arctic Ocean warmed. They ate a far lower proportion of 'typical' Arctic species, such as polar shrimp and Arctic cod, and replaced them with a higher proportion of capelin – a small fish that is more common in the northern Atlantic, where the water temperature is slightly higher. The capelin is known for shifting its range and distribution very quickly in response to changes in water temperature and has been found to be moving further northward as the Arctic Ocean warms.

But we are unlikely to find out what that new equilibrium could be as sadly the warming cannot just stop today. Even with a global effort far exceeding what we have seen to date, there are changes locked into the global climate system that now seem unstoppable. Multi-year sea ice – ice that did not melt during summer and so becomes thicker in winter – has more than halved since 2002, and what we know of the albedo feedback loop suggests it is likely that the Arctic Ocean will experience sea-ice free summers by 2050 or even sooner, for the first time since human civilisations began. The melt of permafrost, changes in rainfall and wind patterns, disruption of the protective halocline layer through ocean mixing, and alterations in ocean currents are all complex and interlinked variables. Many changes are already here, many more are coming, but it is hard to predict how they will play out.

We cannot know all the consequences for either the Arctic Ocean or the world. Some species will benefit, at least initially, but many will suffer. What is certain is that the delicate, intricate balance of this unique, mesmerising and complex marine environment is already lost. But if we redouble the speed and scale of our efforts we may just be able to stabilise it enough for a new balance to be established one day.

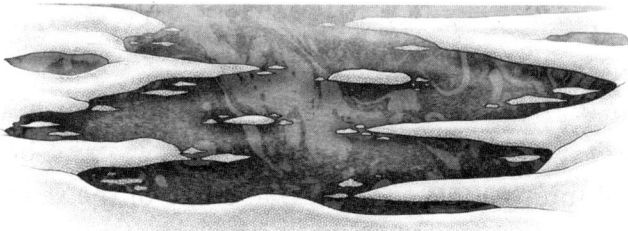

THE LAST ICE AREA

It will be the Holocene's last stand, a remnant recalling what 'stable' and 'normal' once looked like. When the summer sea ice has disappeared across the Arctic, this will be the final piece to go. The Last Ice Area.

We are situated on a coastal fringe in the high Arctic. Spanning a spectacular group of islands, bays and wide flooded valleys known as sounds, where the far northern extent of Canada meets Greenland, on the edge of Baffin Bay, this will be a refuge for ice-dependent species and the Inuit people who have lived sustainably here for thousands of years.

Even here, the ice has still been melting – albeit not quite as fast as in other parts of the Arctic. But while out of this loss came a threat, out of the threat has come something wonderful. The melting ice has made this region more accessible and therefore economically viable for both shipping operations and companies in search of more of the same fossil fuels that drove the loss of ice in the first place. However, these increased pressures provided the impetus and urgency for the creation of a visionary project – the immense Tallurutiup Imanga National Marine Conservation Area.

Tallurut is the Inuktitut name for Devon Island, a central feature of the conservation area, and it got its name because the local Inuit people thought that some of the streaks on the land of Devon Island resemble a facial tattoo on a jawline. *Imanga* means body of water. Tallurutiup Imanga: the name both applies to the body of water between Baffin Island and Devon Island and the entire marine conservation area, which is even larger. But it is not just the name that originates from the region's indigenous peoples. Too often, around the world, the creation of protected areas has ignored indigenous peoples' knowledge and wishes, and sometimes has even taken their territory and abused their rights. The foundation of Tallurutiup Imanga came from a very different vision. In 2019 it was established as an agreement between the Qikiqtani Inuit Association, which represents over 15,000 Inuit, and the Canadian government. Crucially, at its heart was an Inuit impact and benefit agreement. To Steven Lonsdale, an Inuk born and raised in Iqaluit and now director of Marine and Wildlife at the association, this is what makes Tallurutiup Imanga special. He says:

Tallurutiup Imanga is a park, but it is not your typical park such as you would find anywhere else in Canada. It is used in a very different way. In any other part of Canada, a park is set up and geared towards visitors and how to best maximise the visitor experience. But here it is set up as a way to protect the ocean and land together with the cultural activities and cultural continuity of the people who are already there.

Tallurutiup Imanga could have just been a small test of this approach, but instead it has been done at a supersized scale. It is one of Canada's largest bodies of protected water and illustrates a model that could transform marine protection everywhere.

The park takes superlatives and engulfs them. It is not merely 'vast', 'immense' or 'epic', as a guidebook might say – at 108,000 square kilometres it could contain nearly all of England; or, to use a more local example, it is the size of

Nova Scotia and New Brunswick combined. To say it is 'rich in wildlife' fails to do justice to the sheer abundance of walruses, polar bears, bowhead whales, belugas, bearded seals and migrating birds that frequent it. To stress it is of 'global ecological importance' still doesn't quite convey what it means to be one of the last places on Earth where a functioning Arctic marine ecosystem may endure for another generation. Similarly to simply describe it as 'home' to five Inuit High Arctic communities cannot capture the deep culture and unique way of life resulting from millennia of continued living in the region. This is a truly special place.

The Arctic Bay, Clyde River, Grise Fiord, Pond Inlet and Resolute Bay Inuit communities of Tallurutiup Imanga all depend upon movement. Although they have settlements, important parts of their existence are nomadic, requiring freedom of movement over a large area to find food, gather supplies and for cultural practices. For example, while there are some berries and plants that can be gathered in summer, this is not a climate that can be farmed. Consequently, much of the Inuit diet relies on harvesting sea mammals, birds and other wildlife that are transient, thus requiring hunters to travel over great distances to obtain enough food for their communities. The ocean functions as their highway – albeit one of ice in winter and water in summer – and Tallurutiup Imanga is the main artery that connects the region.

Movement and connection are also important for the wildlife of Tallurutiup Imanga. Its nickname, the Arctic Serengeti, comes partly from the abundance of wildlife but is also a nod to the major migratory corridor from Baffin

Bay, in the east, to the Arctic Archipelago, in the west. In connecting these winter and summer areas it is a crucial refuge and refuelling zone for Arctic species, including for an astonishing 70 per cent of the world's narwhal, 20 per cent of Canadian beluga, the largest subpopulation of polar bear in Canada and some of the biggest seabird colonies in the Arctic. The nutrient upwellings are a key reason for so much life within Tallurutiup Imanga, driving high productivity inside the marine conservation area and nourishing an even wider expanse via powerful currents. A compelling motive for creating an Arctic protected area on such a scale is that the environment itself requires scale to function – whether that's Inuit hunters searching for transient food, polar bears prowling a vast range or migrating species following the seasons. For life, if it is to thrive in such a place, needs space to move.

Travelling over sea ice in temperatures of minus 30 degrees Celsius, navigating treacherous waters with complex currents, understanding the behaviour of wildlife and passing on cultural knowledge and local expertise are all skills that abound within the Inuit communities of Tallurutiup Imanga. Those same communities are also in need of jobs, income and food security. There was an opportunity to meet the needs of both people and wildlife. For Tallurutiup Imanga National Marine Conservation Area to be effective in the long term, it needed to be designed in such a way that the communities would benefit from the protection of the environment and the environment would benefit from the long-term stewardship of the community.

The Inuit culture is particularly well suited to such a

model, as Steven Lonsdale explains. 'The protection of narwhal, seals or other wildlife is the protection of the culture,' he says. 'Inuit don't really see a distinction between culture and environment – it is one and the same – we are part of the environment and it is through the cultural connections that both are preserved.'

He also points out that while Inuit have been pushing for protection of Tallurutiup Imanga since the 1970s, its eventual creation has come at a particularly opportune moment when many Inuit in the region now spend part, or all, of their lives in permanent communities.

If you are in an urban centre it is easier to think you are apart from nature or above it in some way, but coming from a hunter-gatherer society that only one generation ago lived a semi-nomadic lifestyle, the connections are still there. My grandmother and my mother lived that way, living by the seasons which dictated where you went. I'm the first who was born and raised in a town. They were so in tune with the environment they could know to go to one place in one weather condition, another during a different season. How successful you would be in terms of food and health was dependent on following the seasons, understanding the ice and the wildlife.

We are still close enough to that generation and that culture that setting up the park in a way that has cultural continuity means we can be the stewards of the environment and the environment remains the centre of our culture – in fact the environment can be actively used to regain and revitalise our culture.

To achieve this the Qikiqtani Inuit Association set up the Nauttiqsuqtiit Stewardship Programme, which was designed to interweave the success of Tallurutiup Imanga with benefits to the community. The programme applied specialised skills that the Inuit uniquely possessed, including tracking, travelling over ice and an in-depth understanding of the hazardous currents and weather, with key functions needed within the conservation area. The Nauttiqsuqtiit, or Inuit stewards, are described as the eyes and ears of the marine conservation area. The stewards gather information on the health and abundance of wildlife at locations such as Pond Inlet where there have been concerns about the impact of increased shipping and seismic testing – blasting the sea floor with sound to map for oil and gas. The data they gather will be vital in developing future marine management policies. They also act as cultural liaisons and interpreters, aid search and rescue, monitor the waters for boats entering without permits, and lead cultural activities such as taking young people from the community on week-long fishing camps or older members of the community to visit ancestral lands.

Until he was six years old, Mishak Allurut experienced family life as it had been for generations: sleeping at traditional campsites, travelling as conditions dictated, living off relatives' knowledge of the region and deep community connections. Then they moved to Arctic Bay, settling for the first time in a permanent community so that Mishak and his siblings could attend school. He left at thirteen and took various jobs, including as a mine worker and wildlife officer. His colleagues remember his sharp mind and generous spirit, both of which he would later put to great use.

Mishak returned to education as a mature student, combining his studies with expert observation of the ice to publish research papers assessing climate vulnerability in Inuit communities. A natural curiosity and interest in people led him to become a connector of worlds, passionately championing the blending of science and Inuit culture. He saw the combination of the two as key to understanding and protecting a changing Arctic.

In the latter part of his career Mishak worked within Tallurutiup Imanga as a Nauttiqsuqtiit coordinator and translator. Fluent in both Inuktitut and English, and knowledgeable about both science and culture, he played a key role in bringing together the different expertise and interests required to set up the park. Where others saw barriers between cultures, Mishak saw a common interest, commenting: 'Tallurutiup Imanga is a project for the whole country, all of Canada, not just for Inuit. Around the world it is a recognised marine conservation area. We are part of that, we are proud to be part of that.'

But his greatest passion was connecting young people with their traditional culture. Perhaps in an echo of his own dual childhood – part nomad, part high-school student – Mishak and the other Nauttiqsuqtiit stewards took young Inuit out into the park to teach them igloo building, harpoon making and traditional hunting. For Mishak, this meant building a connection with the land and with the ice: 'I know there are youths without fathers. We can provide them the opportunity to learn. We can teach them the traditional way to catch seal with a harpoon. It gives them the skills to think and understand the land.'

Before he died from cancer, on 13 July 2023, Mishak passed on his passion, knowledge and culture to his community. There is a plaque in the translators' booth dedicated to him and his work – an everyday hero honoured by the community he served. In the picture beside the plaque Mishak is standing by the edge of the sea, the sun behind him and the summer ice floating on the water. The summer sea ice may not endure until his seven children, fifteen grandchildren and one great-grandchild reach the age Mishak is in this picture, but thanks to his work, and that of his colleagues, many other aspects of the culture and environment that makes Tallurutiup Imanga so special will last.

The Inuit stewards have another vital role. Food insecurity in the north is extremely prevalent – grocery prices are far higher than in a mainland Canadian city and wages and employment levels are low. All groceries have to be imported by air or sea, which is both expensive and polluting, and these are not the foods which make up the local traditional diet. Part of the Qikiqtani Inuit Association plan was to reduce food insecurity by supporting a traditional diet and food culture, and the stewards have a vital role to play in harvesting marine animals to share with their community. Food sharing is an important part of Inuit culture: it is a sign of respect and a cultural bond as well as a practical solution to living in a part of the world that often requires long journeys to find enough to eat. It is far less risky for a small number of expert hunters to navigate the treacherous conditions and bring back enough food for the community than it is for everyone to hunt. The practicalities of such a diet also ensure it is sustainable. Human population density is low and the hunters harvest a relatively small number of marine mammals, fish and birds from a particular location. The next time they hunt the animals themselves might have moved on or changing sea and ice conditions will dictate that the hunters must go to a different location, so the impact is spread. Indeed, as climate change accelerates, there is every likelihood that the Inuit will have to travel further and further in more unstable conditions to find food and supplies.

The Inuit have inhabited Tallurutiup Imanga for thousands of years and they have been calling for its protection for decades. Now it seems that the political will has caught up. The need to conserve biodiversity and tackle climate change has led some in the western world to recognise the core value of Inuit culture: we are a part of nature, not separate from it or above it. Tallurutiup Imanga is a vision of ocean conservation that places indigenous peoples at the centre of the design, management and purpose of a protected area, and it is this that gives it the greatest chance of long-term success.

And the approach is spreading. If Tallurutiup Imanga is on the southern edge of the Last Ice Area, then Tuvaijuittuq is right at its centre. Meaning 'the place where the ice never melts' because it is the location of the oldest and thickest multi-year pack ice in the whole of the Arctic, Tuvaijuittuq is an essential summer habitat for ice-dependent species such as walrus and polar bear. In the Arctic Ocean, off the north-west coast of Ellesmere Island, Tuvaijuittuq is a further collaboration between the Qikiqtani Inuit Association and the government of Canada. This is a marine conservation area of 319,000 square kilometres, over three times the size of Tallurutiup Imanga – and the two protected ocean areas combined are larger than the whole of Germany.

Tallurutiup Imanga and Tuvaijuittuq have a tricky future to navigate. Even if the world does now decarbonise rapidly, our emissions to date are sufficient to cause a lot more ice to melt in both areas. This will bring more demands from the outside world. Some, such as tourism and fishing, could generate valuable income and may form part of a long-term management plan, though there could be pressure to expand both beyond sustainable levels. Others, such as new mining and fossil fuel operations, will hopefully remain excluded. The ice melt will inevitably change nutrient flows and wildlife migrations in ways we cannot yet predict, impacting the entire ecosystem including the lives of Inuit communities. But by taking the visionary step of creating vast, community-led marine protected areas, the Qikiqtani Inuit Association and government of Canada have given themselves the best possible chance of sustaining the unique culture and wildlife of the Last Ice Area.

6

MANGROVES

CAPUCHIN MONKEYS, COSTA RICA

Mangrove forests hang between two worlds. Viewed from above, the dense canopy would not suggest this is a vital *ocean* habitat. But each day the rising tide floods the roots, trunks and sometimes even the canopy, so the forest becomes a place of shelter for the young of big fish and rich waters for an abundance of shellfish to filter feed.

Only a few of the trees in these forests actually belong to the mangrove plant genus *Rhizophora*, and what we refer to as mangrove forests may contain a number of different species of tree, all of which have adaptations that enable them to thrive in salty waters along the coastlines and river estuaries of the tropics – conditions which would kill most plants. They are extraordinary and unique habitats. Such mangrove forests are home to many kinds of insects, birds and crustaceans, but it takes a special kind of primate to exploit them – one that can behave in tune with the tide.

In 2001, while making *The Life of Mammals*, I spent several weeks in one of these forests hoping to film the behaviour of two separate troops of extremely intelligent monkeys – capuchins. We wanted to begin the programme by showing how sheer intelligence enabled different monkey species to thrive in a variety of difficult conditions – it is fair to say that none of us expected to be filming it in an ocean habitat! But we had read scientific studies of the way capuchins harvested shellfish at low tide in a mangrove forest in Costa Rica, and it sounded an interesting way to begin the programme.

The capuchin is a particularly clever species of monkey. Capuchins are often described simply as 'inquisitive', but when watching them at close range for a period of time you realise that, much like ourselves, they are able to imagine the future and plan how to deal with the problems it will bring – exactly, as it turns out, the characteristics required to exploit the complex world of mangroves.

Our filming location was a former cattle ranch and mango plantation that, as part of Costa Rica's admirable forty-year push to restore its lost forests and wildlife, was now well on its way to becoming a fully-fledged wildlife reserve. Some twenty years before our visit the area had received protected status from the Costa Rican government in order to look after its mangroves, and already it was thriving. The mangroves fringed a sheltered bay, and every day, as we walked from the entrance through the semi-deciduous woodland and out to the mudflats of the mangrove forest, the creatures around us were so abundant, beautiful and fascinating that we had to keep reminding ourselves of why we were there and focus on our goal.

Orange-fronted parakeets and yellow-naped parrots were fairly common. Occasionally we would see the instantly recognisable silhouette of recently reintroduced scarlet macaws flying above or catch a glimpse of a manakin through the trees. Iguanas rested in a statuesque way at the end of branches as if reaching out for the sun's warmth, and the roars and grunts of howler monkeys never failed to conjure up the feeling of being observed by a far more terrifying creature than the actual noise-makers. While this was not pure, untouched wilderness, it was clear that nature was regenerating.

Capuchin monkeys are so named because of their perceived resemblance to the religious figures of the same name who wear dark brown robes and large hoods with only their paler faces visible. The monkeys are medium-sized, with adult males growing up to half a metre in length and weighing 3 to 4 kilograms. Their long grasping tails support them when travelling through the trees and they typically live in troops of between ten and thirty individuals. Interestingly, both troops in this mangrove forest were towards the upper end of that range, perhaps indicating the nutritional value of the diverse food sources available.

For me, the first revelation was that the capuchins were well aware of the rhythm of the tides. How they knew or understood this I cannot say, but every day, within ninety minutes of low tide, there they were, out on the newly exposed mudflats. Local scientists had tracked them for many months and discovered that they had a well-established circuit taking in good fruiting trees, suitable areas for leaves and grubs and, at the right moment of the tide, marine protein was clearly on the menu.

The unique nature of mangrove forests makes them

especially good places for shellfish such as clam and oyster to grow. Because the ocean water mixes with the rich mud and plant debris every high tide, it is particularly thick with nutrients. This is in part what sustains so much ocean life in a habitat that, in effect, only exists for half of each day. Filter feeders like clam and oyster obtain their food by washing large quantities of water through their gills every day, the particles sticking to the mucus inside their gills. This mangrove forest was well established and so the water was ideal for clams to live. But unlike the baby fish that would float out of the forest when the tide retreated, most of the clams would remain, often buried in the mud. And these clever monkeys knew this.

In many other sites across the capuchins' range, scientists have documented how they find food in places that other species might well ignore. This is almost certainly due to their good memory and imagination. They look for food everywhere. Their vision is excellent but their sense of smell is poor, so to locate food that is out of sight they use their brains. Unlike most animals, they are seemingly able to imagine that there might be food underneath a leaf or rock or inside a hole in a tree, and then search for it.

We couldn't hope to track the capuchins in the mangroves; they moved through the tangles of aerial roots much faster than we could. But we found a suitably open area well stocked with crabs, clams and oysters, and waited. Eventually a troop of capuchins arrived. Some of the braver ones plunged their hands into holes in the mud. The successful ones pulled out crabs, the unsuccessful quickly withdrew in pain! It was fascinating to watch. But the behaviour we really wanted to film was the way in which they located and ate clams.

They moved with the ebb and flow of the tides. Each day the muddy ground would be exposed approximately fifty minutes later than the previous day, and the monkeys adjusted their movements accordingly. By the time we had been filming for a few days, they took little notice of us and allowed us to get close and film as they dug in the mud and located clams.

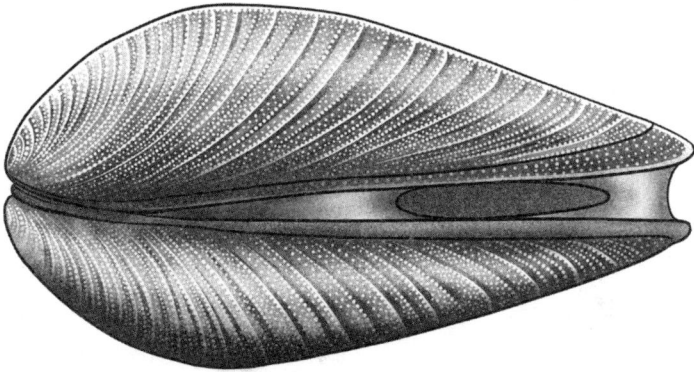

Relative to their size clams are enormously strong, and opening them requires ingenuity, not force.

The shellfish clamp the two halves of their shell so tightly that even a human can't open them without a knife or similar tool. But the capuchins have worked out their own way of getting at a clam's flesh. Having collected one, they take it to a convenient branch and start knocking it, over and over again. Eventually the clam gets so tired it relaxes its muscle and the capuchin is able to prise it open. Younger monkeys learn this technique by watching the older ones intently, and then practising again and again. Some never quite get the knack but most do, for the behaviour has persisted for generation after generation.

But not all clams are equal. Sometimes it took so long to exhaust a clam's strength that the tide had returned before the monkey was successful. Nonetheless the monkey seemed well aware of how much effort it had already invested and that perseverance would be rewarded, for when it was time for the mangrove forest to become part of the sea once again, it held on to its victim as it retreated in front of the rising tide, bashing the shell on the mangrove roots as it went.

DA

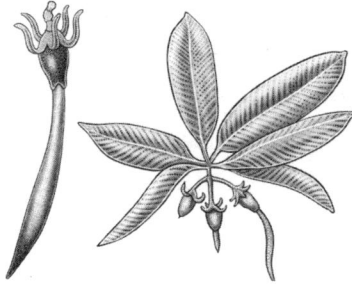

BETWEEN TWO WORLDS

There is a place where sharks swim over the footprints of tigers: two top predators hunting in the same tangled forest separated by the orbit of the moon and the tide that it pulls across the muddy land. Where the mighty waters of the

Ganges, Brahmaputra and Meghna form the largest delta on Earth, flooding into the Bay of Bengal, lies the Sundarbans – a collection of low-lying islands around which grows over 10,000 square kilometres of mangrove forest.

Mangrove forests span two very different realms. At low tide trees, often with long exposed roots, sprout from muddy or sandy soil forming an intertidal forest one could stand in. Yet every single day, when the tide rises, the ground becomes ocean. The plants that live here are an evolutionary miracle, able to thrive despite some of the most extreme daily changes of any habitat on Earth.

The Sundarbans is one of the true wonders of the natural world, home to many globally threatened marine species such as finless porpoise, Irawaddy dolphin and olive ridley turtle, alongside fishing cats and smooth-coated otters. Hundreds of species of birds feast on the abundant insect populations, small fish and fruiting plants while a multitude of crabs act like thousands of ecosystem engineers – eating leaves, recycling nutrients, burrowing and reworking the soil and sediments, oxygenating it in the process. Even more plentiful but far less revered are copepods, a kind of tiny crustacean that drifts in and out with the tides, adrift in the water column, too small to direct their own travel. They thrive in the warm brackish waters; indeed, in certain places there are so many of them densely packed together that the water itself appears almost smoky. They may not be the wild creatures most of us imagine when we picture the Sundarbans, but these tiny copepods have had a disproportionate impact on our lives.

For thousands of years both local communities and the ruling emperors had left the Sundarbans well alone despite the attraction of its fish populations. Daily flooding, malaria and hard-to-navigate, spiky forests created conditions far too hostile to bother with. But in the late eighteenth century the British East India Company had other ideas. Over just a few decades it slashed hundreds of square kilometres of mangroves with a view to developing agriculture. Banks were built, waterways diverted and settlements established. As a result, the 1-millimetre-long copepods began to find their way into people's bodies – just a few droplets of water, perhaps ingested when splashing one's face to cool down, would be enough for them to infect a person. The copepods themselves were not the problem. The issue was the vibrio bacteria that collects in and on them. Initially the bacteria had little impact but, in time, repeated exposure allowed the bacteria to adapt to infect its human hosts. It had become a zoonosis: a disease that could be transferred from animals to humans.

Infection would doubtless have been a horrible experience. Many would have died. But beyond the Sundarbans it was not a cause for concern. Then in 1817 the vibrio bacteria evolved to such a point that it could pass directly from human to human without needing its tiny crustacean host reservoir. It could spread, and spread fast. Within just a few months *Vibrio cholerae* swept fear and death across thousands of kilometres. Not long after, cholera colonised much of the world. A single outbreak quickly turned into the world's first pandemic.

Mangroves are home to a host of fascinating creatures but it is the trees themselves that are the true wonder. While it is common to speak of mangroves as if they were a single plant or plant family, mangrove forests consist of a variety of trees and shrubs that can grow in this environment. An interesting feature of mangrove systems is that any one forest will typically have very few tree species but will support a rich variety of wildlife species. Indeed, some of the most biodiverse mangrove forests contain fewer than five species of tree. This is in contrast to other types of forest at the same latitudes, which will generally have a wide variety of tree species.

Depending on how you classify them, there are globally between seventy and eighty species of tree that make up mangrove forests; they belong to different taxonomic groups but are similar in the adaptations they have developed to enable them to live in such challenging conditions. Together they form one of the most important habitats on the planet, equally vital to life on land as they are to life in the water.

Mangrove forests typically grow in the tropics, in conditions so salty they would kill most other tree species. Almost all those that do thrive do so because of the ingenious ways in which they have evolved to deal with the salt in the water.

If you ever find yourself so motivated as to lick a mangrove leaf – not advised as some are toxic – you might find it has a salty taste. This is because many types of mangrove tree extract the fresh water from the seawater and get rid of the salt by secreting it through glands in their leaves. Other species filter out the salt as it enters their roots, and a few have even evolved the survival tactic of directing the salt to their bark or to older leaves. As the bark and leaves are shed, the salt goes with them.

Many mangrove trees have aerial roots – roots which are still anchored but largely above the ground. They take a variety of forms and can be stilt-like, pencil-like, ribbon-like, knee-like, peg-like or even mushroom-like. In many cases these are breathing roots with pores that allow oxygen to enter the root system and compensate for the oxygen-poor soil they grow in. For some trees their aerial roots serve a further function – broadening the base of the tree and supporting the main stem against the force of the tide. The tangle of roots also slows down the water as it moves through the forest, causing the suspended silt to settle; they then help bind together the sediment into thick mud which can be several metres deep and stores enormous amounts of carbon.

Mangroves create a dynamic system. Even without human interference they will expand in some areas and contract in others. The mud their roots collect can, over time, create land. With every tide a little more sediment is added and the layer of mud thickens until it eventually becomes terra firma. Other species of plant move in to replace the mangroves and a new bit of terrestrial forest is created where the mangroves once stood.

Some coastal wetlands contain both saltmarsh (a salt-tolerant ecosystem with grasses, sedges and herbaceous plants) and mangroves, and, depending on conditions such as temperature, one ecosystem will out-compete the other.

In this manner coastal wetlands can naturally transition between mudflats, mangroves and terrestrial forest. So not all loss of mangrove forests is caused by humans and not all gain is from restoration. That said, while records are imperfect, it is estimated that approximately 50 per cent of mangroves have been destroyed by humans in the past century and half of the remaining mangrove ecosystems are classified as 'at risk of collapse' by the International Union for Conservation of Nature Red List of Ecosystems.

Advances in satellite monitoring mean we have a much clearer picture of the extent of mangrove forests today. About 15 per cent of the world's coastlines are covered by mangroves, spanning a total area of nearly 150,000 square kilometres, with the greatest concentrations in Indonesia, Brazil, Australia, Mexico, Nigeria, Malaysia, Myanmar, Bangladesh, Cuba and Papua New Guinea. Since 1996 we have lost 11,700 square kilometres and gained 6,455 square kilometres. In just the last ten years the rate of deforestation has fallen dramatically, and according to the United Nations over 40 per cent of the remaining mangroves in the world are now protected.

The other key characteristic of a mangrove forest is created by the ocean – specifically the tide. Some locations have a very small tidal range, as little as a single metre, while others can be much larger, resulting in very different forests. Most mangrove forests are flooded twice a day, though some are in diurnal locations with just one high and one low tide per day. In this way mangroves can feel like not just one incredibly rich habitat but two – marine and terrestrial – with two distinct casts of characters that take their place in these forests, crossing at the curtain call of the tide.

As a marine habitat, a mangrove forest's slow-moving water and tangle of roots makes an excellent nursery. Because most mangroves are in the tropics, sometimes there are coral reefs nearby. Recent research has shown that where this is the case, the survival of the mangrove forest is vital for the survival of the coral reef, as important reef species spend their juvenile stage in the mangroves, only leaving to return to the reef when they are fully grown. At over a metre long, rainbow parrotfish are the biggest herbivorous fish in the Caribbean. As such they are crucial to the health of the reefs in that region as they spend most of their day eating algae which would otherwise grow over the corals. Studies have found that the juveniles are highly dependent on mangroves, and that as a result there are very few adults on the reefs in the Caribbean that do not have nearby mangroves. Juvenile turtles and sharks also favour mangroves for the shelter they provide, the small gaps in the tangle of roots being too difficult for larger marine predators to penetrate.

But mangroves are not just for youngsters. In El Salvador, fishers had told stories of critically endangered hawksbill turtles nesting in the dry land among mangrove estuaries. This was not just considered to be highly unusual behaviour but extremely unlikely, as the hawksbill had been thought nearly extinct in this region. But the fishers were right. The discovery generated great excitement, revealing not only more evidence of the value of mangroves but also the value of local knowledge and expertise. Several such nesting areas have now been discovered and offer new hope for this species.

Bivalves such as clam and oyster are present throughout the tidal range. As filter feeders they thrive in nutrient-rich waters with a high sediment content and so are plentiful throughout all mangrove forests. Indeed, it is quite common to see mangrove roots covered with clusters of oysters or to find large numbers of clams buried in the mud or attached to rocks and wood.

As the tide withdraws and the mangrove forest once again becomes land, a new assemblage arrives. Because the tide times change each day by 50 minutes – the tidal lunar day being 24 hours and 50 minutes – somehow many of the species that arrive at the mangroves must adjust their behaviour patterns to exploit the resources that low tide presents. Fishing bats leave their roosts to swoop and, as their name suggests, catch fish. Flamingos forage for crustaceans and insects, tree crabs descend from higher branches to scavenge in the tree roots, deer such as chital in India or white-tailed in Florida arrive to browse. And in the Sundarbans the Bengal tiger follows, stalking its main prey of chital and wild boar – although, in a behavioural adaptation to exploit this most unusual of tiger forests, they have also been observed eating crabs and fish.

But possibly the most surprising of all is the crab-eating frog. It is, of course, quite surprising to find a frog eating crabs, but the really surprising part is where it is eating them. Amphibians rarely live in salt water because they have permeable skins which make it difficult for them to regulate their internal fluid levels. Placing one in salty sea water would cause these fluids to be drawn out of its body, leading it to dry out and die fairly swiftly. But the crab-eating frog of South East Asia has a remarkable solution. Exactly how it achieves this is still a mystery, but it can retain its urea – the waste product many animals produce and the major component of human urine – and by doing so, the internal fluids in the frog's body are sufficiently similar to salt water for it to live quite happily in its mangrove habitat. What makes this neat trick even more remarkable is that the urea in the frog's body reaches levels high enough to be lethal in most other creatures. This inventive but confounding adaptation means the frog can exploit the abundant resources of the mangroves when many of its competitors cannot.

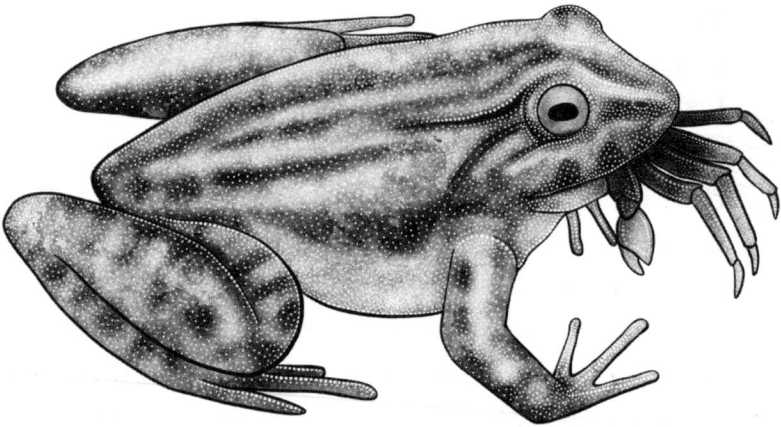

A crab-eating frog, doing exactly as its name would suggest.

Mangrove forests are important food sources for us as well. Millions of fishers rely on them both as locations for gathering seafood and because many of the fish and crustaceans they catch at sea grew up in the mangroves. Clams, oysters, crustaceans, fish and even honey are widely harvested from mangroves, and such is the richness of these forests that relatively large volumes can be gathered by communities in a sustainable manner. Indeed, a whole food culture and folklore has evolved in parts of South America around the mangrove-frequenting *concha negra* (black clam) and *cangrejo rojo* (red crab).

But perhaps mangroves' greatest value to us is the protection they provide against flooding and climate change. There is clear evidence that the devastating impacts of the Boxing Day Indian Ocean tsunami in 2004 were lessened in areas with healthy stands of mangroves, sparing countless lives and homes. Less extreme flood events are also reduced by the densely intertwined mangrove forests breaking up, slowing down and dissipating storm surges. In this way mangrove forests protect millions of homes and reduce property damage by an estimated 65–82 billion dollars every year, simply by being there. And they require no maintenance – in fact they replenish and extend their range easily and efficiently if left to do so. With climate change heating our ocean, raising sea levels and fuelling more intense and frequent storms, we need the return of mangroves as fast as possible.

Mangroves can also mitigate runaway climate change itself. Their role in carbon storage and sequestration is of global significance. On average, 1 hectare (10,000 square metres) of mangrove forest holds some 907 tons of carbon. By comparison, boreal forest stores 350 tons per hectare, temperate forest 340 tons per hectare and tropical upland forest less than 300 tons per hectare. Again it is their dual nature – both marine and terrestrial – that makes them so efficient at gathering and storing carbon. Like any other forest, the growing of trees and shrubs draws down carbon dioxide from the atmosphere during photosynthesis. It is absorbed through the leaves, the carbon is used and stored in the growing wood, and the oxygen 'waste' then released. In mangrove forests dead roots, trunks and leaves are often buried and remain stored in the low-oxygen soil, locking away carbon. Additionally, their ingenious root structures can also capture carbon from *outside* their own ecosystem. Their elevated roots slow down water flow and form perfect nets to capture pieces of seaweed and seagrass that wash in on the tide, which, rather than rotting on the shore and releasing their stored carbon, instead settle with the silt into a rich mud teeming with crabs, worms, bacteria, nematodes and many others, breaking down organic matter and forming one of the most carbon-rich soils on Earth.

The richness and biodiversity of the mangroves further increases their carbon storage. Even fish faeces adds to this great carbon super-sink. Many fish capture carbon when feeding on plant life such as algae; they defecate and then, as the tide withdraws, their waste sinks into the mud and acts as fertiliser, which in turn helps the trees to grow, drawing down even more carbon from the atmosphere in the process. Furthermore, because of the conditions in

A mangrove forest in the United Arab Emirates. This aerial view shows mangroves' ability to thrive in salty conditions, enabling them to provide vital habitats for both marine and terrestrial species as well as flood protection for coastal settlements.

Two juvenile lemon sharks cruising the edge of a mangrove forest in the Bahamas. These mangroves were probably the sharks' nursery and offered protection during the first stage of

their life. Now, as juveniles, too large to swim through the tangled roots, they return regularly to feed in the deeper water and wider channels between the plants.

A mudskipper in the Sundarbans mangrove forest. As an amphibious fish, able to 'walk' on land and also absorb oxygen from the water, the mudskipper is ideally adapted for the dual world of mangroves. They have even been observed climbing the roots of the mangrove plants.

the mud beneath mangrove forests – low oxygen, salty, waterlogged – breakdown of organic carbon can be very slow, which means that carbon is stored for much longer. In mangroves in Belize, it was found that deep layers of mangrove mud had turned into peat and that the deepest layers, an astonishing 10 metres from the surface, were more than 6,000 years old.

Our world needs its mangroves. The more of them we can protect and the more we can allow to regrow, the better. In some places it will be too late for the mangroves to return – their former range concreted over for homes, ports and other developments. But thankfully in many locations where mangroves formerly existed recovery is possible if the right incentive exists. Indeed, a global study in 2018 identified 8,120 square kilometres of potential mangrove habitat as suitable for restoration.

On occasion it will be necessary for local communities or conservation organisations to actively begin this process, perhaps by removing barriers to the tide or creating channels to allow seawater to recreate marine coastal conditions, thus preparing it for mangroves and killing off competitor plants. But the complex biology of mangroves makes them ideally suited for natural recolonisation. Their environment is by its very nature dynamic – for example, changes in sediment washed down through rivers can lead to an area transitioning between wetland, mudflat and mangrove ecosystems. As a result, the tree species that thrive in mangrove forests are also adept at colonising new potential habitats when the opportunity arises.

Furthermore, some mangrove seeds have ingenious

characteristics to help them spread. They germinate when still on the tree, often growing into seedlings while still attached, so they are ready to take root the moment they drop. Some mangrove propagules are torpedo-shaped with a stem up to 70 centimetres long so that when they fall they can plunge directly into the mud – immediately ready to grow. And since there is a roughly even chance that they will fall into water, many can also float, enabling them to travel until they reach water shallow enough for them to lodge in the mud and grow. Of course not all will be fortunate enough to make landfall in perfect mangrove habitat, but enough do to make this an evolutionary benefit to the tree.

At this point in human history when there is an urgent need to restore nature, temper global heating and protect ourselves against the violent storms fuelled by the warming we have already created, mangroves are our ideal ally. Wherever we allow them to remain or regrow, they will provide food and flood defences, and enable countless marine species to breed and recover all while helping us draw down carbon and restore balance to our world.

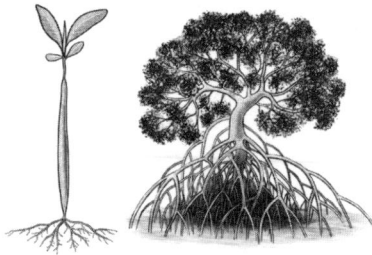

MANGROVES

GULF OF NICOYA, COSTA RICA

Ana Guzman stood on the edge of a sugar cane plantation fringing Costa Rica's Gulf of Nicoya. A marine biologist by training, she was proud of her country's world-renowned nature restoration programme. Over the past forty years its cloud forests, birdlife and ecotourism had become a global showcase for how a country can develop its economy and public services alongside protecting and restoring its natural wonders. *Pura vida*, or 'pure life', had become synonymous with Costa Rica.

The reputation endured because there was ample evidence behind the image. It wasn't hard to find world-class examples of forest restoration, or communities thriving by creating businesses which worked with nature rather than against it. Scientists and wildlife filmmakers gravitated to this small Central American country to monitor and document the astonishing recovery of the wild. On the global stage, former Costa Rican government minister Christiana Figueres had recently brokered the historic Paris Agreement on climate in 2015, in part through her expert diplomacy but surely also reinforced by the legitimacy provided by her home country's example.

Costa Ricans had every right to feel proud of their achievements, but Ana knew that didn't mean the work was done. Beyond the world-leading reserves on the tourist routes, Costa Rica and its citizens were still suffering from mistakes of the past. Here, on the edge of a plantation in the district of Puntarenas, the pure life felt a long way away.

Mangrove forests hadn't received the attention afforded to rainforests or cloud forests. Even in the conservation movement many still spoke as if 'mangroves' were just a single species of tree.

Sitting squarely in the tropical climatic zone favoured by mangrove forests, Costa Rica still had 52,000 hectares of mangrove ecosystems and has identified a further 14,000 hectares as having the potential for restoration – including Puntarenas. Located on the eastern side of the gulf – a 90-kilometre-long inlet that separates the Nicoya Peninsula from the mainland – Puntarenas estuary was ideal mangrove habitat. The tide from the Gulf of Nicoya brings the warm salty water of the Pacific across this low-lying ground twice a day, and the area is only under water at the higher points of the tidal range, creating the perfect balance for mangroves to thrive but other competing vegetation to perish. Everything suggested to Ana that this was prime mangrove country with the perfect conditions for them to take hold, grow and spread. Yet all she could see was devastation.

Mangrove forests in the area had in the recent past protected the land from regular storm surges fuelled by the tropical weather systems of the Pacific; provided nurseries for around 80 per cent of the fish species caught by local people, including croaker, snapper and catfish; and been home to many famous Costa Rican terrestrial and marine wildlife species, including the capuchin monkey, caiman, sloth and even the sea turtle. But now most of the mangroves had disappeared or were in poor health. The water channels had become filled with sediment and the ground was almost 3 metres above sea level. As a result, the daily tides no longer entered. Sugar cane from a nearby plantation had gradually encroached and was now being grown illegally, albeit more through apathy than deliberate intent, on some of the 'reclaimed' land.

Ana had spent her whole career working as a marine biologist, with an expertise in fisheries and conservation. Growing up, she had accompanied her family on regular trips from their home in the capital, San José, to the coast, which ignited a formative fascination for marine life. But when the time came to start university, Ana decided that, rather than specialising too early, she would learn everything she could about the natural world in general, including genetics and botany. Eventually the attraction of the ocean was too strong and she focused her attention on Costa Rica's marine ecosystems. Only later would she come to realise the value of understanding botany when it came to protecting fish.

She loved the purity of science but became fascinated by the way that coastal communities interact with the fish populations on which they depend, and realised there was a vital need for conservation to provide a bridge between the world of data and the reality of life for fishing communities. She studied the ocean but understood that sometimes the best way to protect it was to address the problems where the waves meet the land. And that was how a marine expert found herself standing 2 kilometres inland about to embark on a mission that would dominate years of her life.

Despite its enviable track record in protecting other habitats, Costa Rica had been losing its mangroves. The area of Puntarenas where Ana stood had lost 766 hectares of mangroves over the last sixty years, and every indication was that this would continue. She wondered if by presenting an alternative future, where mangroves were seen as valuable, the community here could create a model for recovery across the Gulf of Nicoya and perhaps for all of Central America.

The Gulf of Nicoya is Costa Rica's largest estuary and one of its most important fishing grounds. Located on the Pacific coast, its 150 kilometres of shoreline is considered to be one of the most critical regions in Central America for shorebirds, with surveys estimating that over 20,000 birds use the region during migration season. In its natural state the northern shoreline largely consists of tidal mudflats and mangroves, but over time it had been developed for agriculture, homes and to a lesser extent for tourism. The water is rich with nutrients, delivered by two large rivers, the Tempisque and Grande de Tárcoles, and the gulf supports commercially significant fish populations and pelagic visitors such as humpback whales. Yet years of scientific study of fish catch in the Gulf of Nicoya had chronicled a steep decline in most species. While some of this was undoubtedly caused by overfishing, marine biologists gradually realised there was another significant factor: the destruction of the mangrove forests where many fish species spawn and where their fry grow, nourished by plentiful food and protected by the shallow waters and tangled roots.

As climate change was increasing both the frequency and intensity of storms, Ana and her colleagues at the Costa Rican branch of Conservation International had begun hearing of communities in Puntarenas and other parts of the Gulf of Nicoya being flooded. There was a distinct pattern – the most impacted coastal communities were in those areas where mangroves had been destroyed. This was clear evidence that the loss of mangroves wasn't just an environmental issue; it was directly affecting the lives and livelihoods of the people of the Gulf of Nicoya. The realisation could be the turning point.

Beginning in 2018, Conservation International had worked with a community over several years to restore a small mangrove forest at Chira Island, in the middle of the Gulf of Nicoya. They planted over 2,500 seedlings and for a relatively small-scale project had generated spectacular results. Witnessing first hand how well suited the region was for restoration, Ana now set her sights on a much more substantial project, namely the largest mangrove restoration project in Costa Rican history – and one of the largest in Central America. She knew that across the world many mangrove restoration projects had failed for two main reasons – lack of scientific study of the region concerned, and therefore understanding of what might thrive there, and lack of community involvement – and was determined to get both right from the start.

For months experts from Conservation International and the Costa Rican government conducted an in-depth scientific study of the mangrove habitat in the Gulf of Nicoya. The team set out to discover how much had been lost, where mangroves had previously grown, the condition of the remaining ones, and what stock of carbon they stored. Crucially, they also worked with local communities to catalogue and quantify the benefits that the mangroves provided. The Costa Rican mangroves created ideal conditions for molluscs, shrimp and crabs – the tides brought abundant food to them and the mud and roots gave shelter. Since these valuable shellfish can be harvested sustainably in reasonably large quantities from a healthy mangrove forest, they provide reliable, year-round food and income for the nearby community.

Once the experts had gathered all the facts, they could devise a plan for the whole of the Gulf of Nicoya and determine where restoration would have the biggest impact. Three sites were selected totalling 600 square hectares: a small but biologically important patch near the mouth of the Seco River where the mangroves had been cut down to create shrimp ponds; an immense area next to the sugar cane plantation near the mouth of the Aranjuez River; and an equally large site at the northern end of the Gulf of Nicoya. All three had the crucial elements that Ana's research had identified as essential for a successful restoration. Every location had previously been a mangrove forest, so had the potential to recover naturally once barriers were removed and the process of regeneration had been started. And all three had communities that had been affected by the loss of their mangrove forests and so were keen to be involved and would provide ongoing protection for the new forests.

Restoring mangrove forests isn't simply a case of planting trees and leaving them to grow – it works best when the trees themselves recolonise an area. But mangroves need very particular conditions if they are to flourish. The soil has to have just the right levels of salt, flooding, humidity and oxygen. The tide and river channels have to be recreated to flow through the land in the way they did originally, and invasive species such as sugar cane, in this instance, have to be eradicated. The very ground itself would have to revive before the forests would regrow.

This vision would need a wide coalition of partners in order to succeed. The government owned the land, and its pioneering approach of treating nature and the economy as part of the same system would be key. SINAC, a government organisation formed to manage the national parks, forestry and wildlife – previously the responsibility of three separate organisations – had been granted the legal powers to bring together the state, civil society and private enterprise for undertaking ambitious projects such as this. It is an approach that many countries now talk about but few have actually made work in the same way as Costa Rica.

It is one thing to have science, government support and a well-resourced conservation organisation behind you; Ana had learned, over many years, that the most important ingredient was a passionate, motivated local community with a strong vested interest in the project. And this is where José Jesús López and Odilie Carrillo become the heroes of the story.

José Jesús remembered running around the mangrove swamp as a small boy. For him and his friends it was a fun place to play: at low tide there were lots of channels to explore, plenty of places to hide in or to provide obstacles for the games they invented. Many years later that childhood knowledge would become a vital tool in restoring the forest.

The scientists from SINAC and Conservation International realised that just replanting this land with mangrove plants would end badly. The conservation world is littered with examples of failed restoration projects because the wrong trees were planted in the wrong position with respect to the tides, the soil conditions weren't right at the time of planting, or even where lack of community involvement meant there was no ongoing monitoring of the project. It had been decades since there was last a natural forest on this land. The water channels, evolved over thousands of years, that would naturally have allowed the seawater to flow in at high tide were now gone and as a result the soil had lost much of its salinity. The mangroves' superpower of growing where few other plants can survive now presented Ana and her colleagues with the biggest challenge.

They knew the first goal was to rehabilitate the environmental conditions of the area. But how? They couldn't even begin to move the enormous volume of sediment that had accumulated leaving the ground 3 metres above sea level, and decades without mangroves meant the soil not only lacked the requisite salinity but was now so compacted that it lacked the adequate circulation of humidity and oxygen.

The only practical way to restore such a vast area was to reconnect it to both the ocean and the remaining healthy mangrove forest and then help nature run its course. If this worked then the waterflows would improve the soil and allow the propagules (the parts of the plant that grow into new ones) from the healthy forest to spread to the newly revived area. It made scientific sense on paper, but given the epic scale on which it would have to be done the team needed to be sure.

The biggest problem was that they didn't understand how the rivers and estuaries had flowed in the past when the entire area was a mangrove forest. There was little trace of the old water channels, so how could they begin to restore them in a way that natural processes, such as tidal flow and rainfall, could then accelerate? Equally important was the need to connect the remaining mangrove forest with the degraded area. If they got it wrong, not only would the restoration fail but it could alter the flow of water into the healthy forest and damage − or even destroy − the remaining mangroves.

This was high-risk restoration on a scale never previously attempted in Costa Rica.

Satellite imagery helped. From high above, some of the traces of the bigger historic water channels could just be identified. Finding older aerial images gave a few more clues, but José Jesús was the real key: he actually remembered where the channels were.

He smiles easily and laughs at the irony that when he was a boy the mangrove forest was just thought of as a 'muddy swampy area of no real use' while here he is now, many decades later, using his childhood memories and his role in the community association to help bring back the 'swamp'. Ana reflects that this is the journey the whole of Costa Rica, and perhaps the world, is having to take. 'We didn't realise the value of nature until we removed it. And now we want it back,' she says.

José Jesús's memories combined with modern satellite imaging to create the map which would restore the treasure. The community, the government and scientists agreed on an approach. They would dig an astonishing 32 kilometres of channels, nearly 10 kilometres of which were in such sensitive and hard-to-reach areas that they would have to be excavated by hand.

Ana recalls: 'It was the women who were there first. They understood the importance of starting from zero, of nourishing back and engaging for the long term.'

She smiles and continues:

It's a completely different mindset from men. The women know they have to invest for the future and that's the crucial thing with mangrove restoration – it is not a result that you are going to see today. But they know if they invest their time, the result will be worth it. It is like looking after their children.

Odilie Carrillo is the leader of the Women's Association for the Conservation of the Gulf of Nicoya. In photos taken during the project she is often wearing a blue sunhat with a smiling cartoon face drawn on it that matches her own cheerful demeanour. She was an inspiration for Ana, constantly asking her when she was coming back and when they could 'do more hectares'. Odilie and her association were the driving force behind 117 hectares of the restoration – over a third of the entire project.

To get to the site where they would spend the day digging water channels by hand, the women had to walk over 2 kilometres through thick mud in temperatures frequently exceeding 30 degrees Celsius while carrying their equipment. Their dedication and endurance day after day was spurred on by confidence that the work would benefit their community – not to mention Odilie's passion and good humour.

When the channels were finally complete, nature took over the hard work. Water was once again flowing in and out of the land and in time the soil became saltier. The saline soil killed off any remaining roots of the invasive sugar cane as well as terrestrial plants that had spread to the area. Crabs washed in on the tide dug into the soil and began the process of oxygenating it. Finally, several years later, the labour of the community and the regenerative power of nature had transformed the land. The first areas were ready for mangroves.

Thanks to their specialised propagules that can establish themselves regardless of whether they fall on land or water, mangroves are extremely efficient colonisers – if the conditions are right. The interwoven roots of mangrove plants catch sediment, creating thick mud in which their propagules can take root. The seedlings of many species can grow at over half a metre a year, so they quickly reach a point when they too can reproduce. Ana, José Jesús and Odilie knew that once they got the conditions right, the rest would largely take care of itself.

Today the restoration is well under way. The community is now excavating smaller water channels to connect the big ones and speed up the natural process of the forest reclaiming its former ground.

The government is seeing the benefits. It is able to put its protection of coastal wetlands, in particular its mangroves, at the heart of its commitment to meeting the Paris Agreement, thus making its climate goals easier to achieve. Now, for the first time, it includes reforestation of the mangroves as part of its Payment for Ecosystem Services scheme, which compensates farmers and communities for conserving forests on their land.

In turn the community is benefiting, from government payments, increased flood protection and the harvesting of food such as shellfish and crabs. In the not too distant future it will also enjoy the recovery of fish stocks and profit from the growth in ecotourism activities such as birdwatching.

And finally, Ana and the conservation science community are benefiting. Suddenly there is a model that works for everyone and critically it is a model that has been proven to work across a vast area. Ana has good reason to hope that this approach will spread across Costa Rica and Central America – and, in time, perhaps even across the rest of the world's potential mangrove habitat. She knows there is urgency, but she has the patience acquired from years of experience: as she has always said about mangrove restoration, 'It's not a project, it's a process.'

7

OCEANIC ISLANDS
AND SEAMOUNTS

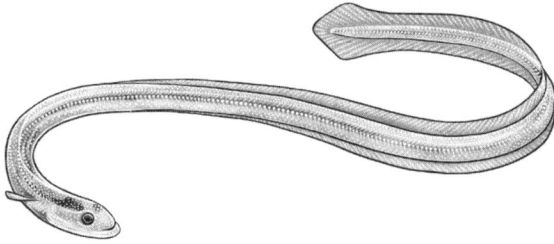

THE TURTLES OF RAINE ISLAND,
THE CORAL SEA BETWEEN AUSTRALIA
AND PAPUA NEW GUINEA

We were supposed to be in the mountainous jungle in the heart of New Guinea. Cameraman Charles Lagus and I had spent a month in the Wahgi Valley in the centre of the island filming birds of paradise and were preparing to accompany a patrol into the largely unexplored valley of the Jimi River that lay to the north. But the guards who were needed to accompany us had been delayed and could not join us for another month. What were we to do?

This was in the mid-1950s. We had no straightforward radio or telephonic contact with my bosses at BBC Television in London. It was up to me. I pondered and quickly reached a decision. It would be absurd to go back to London only to return to New Guinea after three or four weeks. And we were within an hour or so's flight of one of the greatest wonders of the natural world – the Great Barrier Reef. The reef is not, as one might suppose, a single great underwater wall of coral lying a few miles out to sea and running parallel to Australia's eastern coast.

247

It is a tangle of reefs and islands, some of which are mountainous, and many others that are little more than sandbanks. Ever since I was a boy, I had been thrilled by pictures of its multicoloured, infinitely varied colonies of coral and its islands thronged by immense numbers of breeding seabirds. I had always yearned to see this wonder with my own eyes. This was my chance.

I telephoned a friend in Australia. Who could show us around the Great Barrier Reef? My friend knew just the man: an acquaintance named Vince who lived in Cairns, then the most northerly city in Queensland with an international airport. He had built himself a metal boat and now spent all his time in it, exploring the reef. He would be the perfect guide. Two days later we were in Cairns with Vince discussing where we might go.

Vince's suggestion was simple. Why not sail northwards up the reef from Cairns, stopping to investigate any island or reef that particularly attracted us until we reached Raine Island – the reef's northern limit. It was almost 400 miles away. He himself had never gone as far as that, nor indeed did he know anybody who had. But Raine Island was said to have one of the biggest and most varied colonies of breeding seabirds to be found anywhere on the reef. It was also the world's largest breeding site for green turtles, several thousand of which visited every year. Unluckily, the breeding season was almost over, but Vince thought we might find a few late stragglers, and even if we didn't it was worth a visit since Raine was probably unrivalled as a breeding site for seabirds. If we would like to go, he would certainly take us.

The following morning we bought supplies and set off. Day after day we continued northwards, stopping every now and then to explore the wonders of the reefs, crowded with their fantastic stony colonies of coral, swarming with brilliantly coloured fish. And on the fourteenth day we at last sighted Raine Island.

We could be in no doubt that we had found the right place for, incongruously, a tower stood towards the eastern end of the island. It had been built in the middle of the nineteenth century as a beacon for European ships, to indicate the beginning of a route through the tangle of islands and cays that constitute the Great Barrier. And, happily for us, there were clouds of seabirds as numerous and dense as I have ever seen.

The most abundant were two species of tern – the noddy and the white-capped. There were three species of booby – the common, the brown and the red-footed. But, for me, the most impressive and certainly least familiar were the frigates: glossy black, with 6-foot wingspans and long, deeply forked tails. Frigates spend most of their lives far from land over the open ocean, feeding largely by harrying other smaller birds and robbing them of their catch. But eventually the adults come down from the skies, as all birds must, to lay their eggs. Now the males were squabbling with one another over nest sites. Once a male acquired one, he refused to leave it, driving off rivals and simultaneously attracting females by inflating a huge blood-red throat sac and, at the same time, vibrating his immense outstretched wings.

Despite being off season for egg-laying turtles, a few

were still here. On our first walk around the island we saw perhaps twenty curving tracks weaving through the sand hills. I didn't know at the time, but scientists now believe that every year over 60,000 females travel immense distances to get to this one small and remote island. The few here now were just end-of-season stragglers.

Each morning we found several that had dragged themselves far enough inland to be beyond the reach of a high tide and were now digging with powerful swishes of their fore-flippers. Every now and then they swivelled slightly so that the holes they were creating were circular. When one female was down in the sand by about 12 inches, she started to use her hind flippers as well, until finally the top of her shell was virtually level with the surface of the sand. Then, using just one hind flipper at a time, she begins to widen the downward passage to create an egg chamber. The blade of each hind flipper is amazingly mobile and versatile, and she uses it to remove little handfuls of sand. All the time she weeps to clear sand from her eyes. She gasps, making great breathy exhalations, and follows each one with a sudden intake of air, as though she is still in the sea and preparing to make another dive. The nest hole must not be too deep, for the eggs will need the warmth of the sun if they are to develop. But it must, nonetheless, be deep enough to be beyond the reach of predators. Then she lays a hundred or so eggs, fills in the hole and returns to the sea. A single female may repeat this exhausting process half a dozen times during a single breeding season.

*A green turtle laying eggs; only a delicate
balancing act of depth and temperature
will lead to them successfully hatching.*

The sheer quantity of hatchlings that emerge on the beaches of Raine Island in a single season is hard to imagine. But great numbers are essential because only one out of every thousand hatchlings is likely to reach maturity. Within minutes of appearing on the surface of the sand, most are eaten by birds. Those that do reach the water are then attacked by marine predators. Only a tiny minority reach the relative safety of the open ocean.

The island's tower was built by European convicts who used timber salvaged from the wrecks of ships that had foundered here and limestone they quarried from the island itself. The pits they created still remain, and in many of them lay the skeletons of turtles. It seemed as if some females, having finished their laying and seeking a downhill slope that would lead them to the sea, instead travelled down one that led them inland. Some tumbled into pits created by the builders of the tower, and once there, either lost or stuck on their backs and unable to right themselves, had stayed and died.

Their bones littered the whole island. Disconnected and bleached pure white, many had acquired a strange abstract beauty. Some skeletons were still connected as they had been in life and lay on their sides or backs, half buried in the shifting sands.

Today, the wildlife of Raine, with the consent and co-operation of the Wuthathi people within whose traditional territory the island lies, is properly protected. In a world that has become markedly more dangerous for turtles and seabirds than it was when we were there in 1957, this work has become even more vital. A detailed scientific project now monitors the turtles. Fences and landscaping prevent them from falling over the small natural cliffs or into the excavations that were dug many decades ago by the human builders of the tower.

But as on many oceanic islands, today the biggest threats to Raine's turtles originate from far away. Higher tides caused by rising sea levels are now flooding many of their nests every year. And there is an additional and invisible danger.

While it is still in the egg, the sex of a developing turtle is determined by the temperature of the nest in which it lies. High temperatures lead to the development of females, lower ones to males. In recent seasons over 99 per cent of the turtles that have hatched on Raine have been female. In the past this effect was lowered by some nest sites on the island being in shady areas, others being at slightly different depths or by eggs hatching at different times in the season, but because our world is heating up most of the nests on Raine Island are now at a temperature that produces only females.

It seems that a crash of the Raine Island population of green turtles is all but inevitable without a speedy halt to global heating or significant intervention from local conservationists.

DA

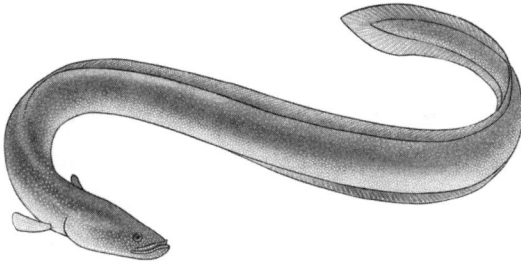

In the remote Pacific, over a day's sailing from the nearest inhabited island, hundreds of hammerhead sharks slowly circle an underwater mountain. Meanwhile, off the coast of New Zealand a dense shoal of fish hover above the summit of an extinct volcano – the water so thick with their bodies that you can hardly make out the deep-sea corals and metre-wide sponges growing beneath them. On the other side of the world twenty-four different species of whale and dolphin all seek out the oceanic islands of the Azores, many appearing to have travelled thousands of miles to reach them. Oceanic islands, and volcanoes and underwater mountains otherwise known as seamounts, are vitally important places for life on Earth that are deliberately sought out by the ocean's wanderers.

But what is a seamount, and why are they such hotspots for marine life? For most terrestrial habitats on our planet, description is fairly straightforward. Advances in satellite imagery and centuries of patient study mean that as far as grasslands, deserts or deciduous forests are concerned we can confidently describe what they are, the approximate area of the planet they cover and broadly how life within them works. There may be quibbles over details and perhaps uncertainty ranges with population sizes, but in essence we have a good idea what we are dealing with. But what is so exciting about seamounts is that we are at a much earlier stage in our journey of discovery.

For now, we consider a seamount to be an independent, stand-alone mountain on the ocean floor which rises at least 1,000 metres; this distinguishes seamounts from the vast underwater mountain ranges such as the 16,500-kilometre-long Mid-Atlantic Ridge. Those mountain ranges are highly significant geological features, just as they are on land, but seamounts operate differently, and are often described as oases in a vast marine desert – one-offs rising from the ocean floor.

However, we are now discovering that they often occur in clusters around oceanic hot spots where the Earth's crust is thinner. This is because most seamounts are, or once were, volcanoes. Scientists have so far located most seamounts either in what are called subduction zones, the points of collision between two tectonic plates, or in sea floor spreading areas where, as the name suggests, two tectonic plates are moving apart. Both locations allow for magma to rise from deep in the Earth and volcanoes to form.

The resulting seamounts can take many forms. Some never reach daylight but others, such as Hawai'i and the Galapagos, break the surface to become oceanic islands. Oceanic islands rise directly from the sea floor of the big ocean basins and differ from continental islands such as Greenland and Tasmania which are simply part of the continental shelf and happen to be unsubmerged at current ocean levels. Oceanic islands may, over time, be broken down by waves and rainfall to sink below the sea once again. Sometimes the erosion creates flat-topped seamounts known as guyots, and on other occasions, under the right circumstances, this process can create one of the most beautiful sights on Earth.

In warm, sunlit waters a volcanic island can provide the rocky substrate coral needs if it is to develop into a reef. Over thousands of years reefs can grow to encircle the island. As the volcanic core of the island subsides, the reefs continue to grow up forming an atoll – a circular reef around a central lagoon. The Maldives in the Indian Ocean and Kiribati in the Pacific are examples of such atolls. Interestingly, Bermuda – often described as the only atoll in the Atlantic – comprises more than 100 islands and numerous reefs in an archipelago around the undersea caldera of a volcano, so very much resembles an atoll and plays a similar function for wildlife, but was formed in a different manner.

Estimates of the number of seamounts vary greatly. We can confidently say there are at least 14,500 that rise 1,000 metres or more above the sea floor. But there are estimates suggesting that in total there may be 100,000 across all the world's ocean basins. As yet we simply don't know.

Low-altitude satellites are able to detect bulges in the ocean surface caused by the bigger, near-surface seamounts – these are the features you can see if you zoom in on the open ocean in Google maps. But to locate the smaller and deeper ones, or to map the true size and depth of a seamount, we require sonar. And, given that we think at least half the world's seamounts are located in the high seas, that typically means a long ocean voyage to use the sonar in situ. It is little wonder, then, that so few have been mapped and so many new ones are still being discovered.

Indeed, in early 2024 the Schmidt Ocean Institute's research vessel *Falkor* discovered four underwater mountains in international waters near Guatemala by using satellites. They first searched for the gravitational pull of the seamount, which draws water towards it and creates slight bulges in the surface height of the ocean. Then, when their boat reached the location the satellite data had identified, they deployed multibeam sonar to create wonderfully detailed maps of the craggy flanks of mountains over 2 kilometres high.

For many decades fishing boats have used sonar to locate

and catch huge numbers of fish on seamounts, from round-nose grenadier and splendid alfonsino to pelagic predators such as tuna. As so many commercially valuable fish are found around seamounts, we have long known that, compared to the surrounding open ocean, they often abound with life. But we don't yet know why. One theory is that because the seamount rises out of an otherwise featureless sea floor it disrupts the flow of ocean currents, diverting deep, nutrient-rich water from the ocean depths in an upwelling. This feeds a bloom in ocean life which assembles in such quantity that it attracts large open-ocean travellers. In effect the seamount creates a hotspot of food for every level of the food chain.

While logically this makes a lot of sense, there are problems with the idea. From the studies conducted so far it appears that upwellings around seamounts don't happen all that frequently, and when they do occur they rarely last more than a few days – nowhere near enough time to sustain permanent local populations of fish in the numbers that have been observed. Then there's the fact that photosynthesis can only occur in the photic zone. So how would the nutrient upwelling theory apply to the deeper seamounts, many of which also appear to attract large numbers of fish?

Some scientists have tried looking at this from the opposite perspective. They have posed the idea that food may be 'trapped' on top of the seamount. The idea is that zooplankton are carried by ocean currents over the top of the seamount. We know that at night zooplankton tend to feed at the ocean surface because that is where the bulk of the phytoplankton they eat is found. Then, in the daytime, they swim to deeper water below 100 metres to avoid predation (some have even been recorded at depths of 800 metres). However, when the zooplankton are drifting over a seamount that is shallower than their night-time depth, they get trapped above it and become easy prey for the fish that eat them. This, of course, means that such fish congregate around the seamounts, attracting species further up the food chain.

As yet only around 300 seamounts have been studied – somewhere between 0.3 per cent and 1 per cent of the total scientists think exist – so in all likelihood this tale has many twists still to come. What we can say is that it seems, for whatever reason, that zooplankton, large invertebrates and big shoals of fish accumulate around seamounts and often this attracts larger travelling ocean species such as shark, ray, dolphin, tuna and whale, many of which appear to deliberately seek out seamounts on their epic migrations.

Indeed, individual seamounts are far more connected to each other than we first thought. The fact that many seamounts, or groups of them, are such vast distances apart originally led to the hypothesis that they are, in biological terms, isolated. This would imply that members of a species on one seamount do not encounter those from another seamount, and that there is no genetic exchange between them. This idea was based on what we know of terrestrial life on islands, where it is common to find species that have spent so many generations on one island that they have adapted to those unique conditions and become distinct from their relatives on other islands or the mainland.

However, while there are some endemic species on seamounts, for the most part life there is connected to the rest of the ocean. For example, larvae can float between seamounts on the ocean currents and adults of many species appear deliberately to travel great distances to reach a particular seamount.

Japanese eels are freshwater fish that spawn in the ocean. Despite the name, they are found across much of East Asia but are best known served in grilled slices under their Japanese name *unagi*. They are a prized fish famed for their nutritional value, cultural heritage and rich taste. But it is their extraordinary life cycle that makes them truly remarkable. While they have been eaten for centuries, their spawning grounds were a mystery until very recently – and it is little wonder this puzzle was so hard to crack. Eels have five distinct stages to their life cycle. Once hatched from eggs they have a larval stage in the open ocean. After about eighteen months they first resemble eels, albeit very small transparent ones known as elvers. They swim towards shore. When they reach about 6 centimetres long they are strong enough to enter estuaries, seemingly compelled by an urge to head upstream. They spend up to ten years in a river, often many kilometres inland, feeding on invertebrates and growing to around half a metre in length. Once they reach full size, their skin turns from grey-brown on top and white underneath to silver. This is the last stage of their life, when they head back to the ocean to spawn. And where are the vital spawning grounds of this important fish? After many years of searching, researchers from the University of Tokyo finally tracked them down. Using genetic analysis, they proved that eel larvae discovered at two seamounts west of the Mariana Islands, approximately halfway between Japan and Papua New Guinea, came from Japanese eels that had migrated over 2,000 kilometres from their freshwater habitat to spawn. Why they choose this spot is still unclear, though it is thought to be connected to ocean currents.

The story of the Japanese eel highlights the importance of seamounts and points to the need to create a new way of protecting both Earth's ocean and its rivers. Japanese eel populations have significantly declined over recent years and the species is now classified as endangered. It is difficult to know how much of this decline is due to warming waters altering ocean currents, fishing damage to seamounts, pollution in rivers or overfishing of the juveniles and pre-spawning adults; the probability is that it is a combination of all these factors. But what the tale of the migrating *unagi* shows us is that if a seamount 2,000 kilometres from land can be important to a freshwater fish then the ocean is a much more interconnected realm than we had previously realised and requires a form of protection that reflects this.

If all the seamounts and oceanic islands were joined together it is thought the area created would be the size of Europe. By that measure alone they would be a significant ocean habitat. But it is possible that by being dispersed they are even more important. Critically endangered scalloped hammerhead sharks migrate between oceanic islands in the Eastern Pacific, aggregating in large schools. One of the least understood marine mammals in the world, the

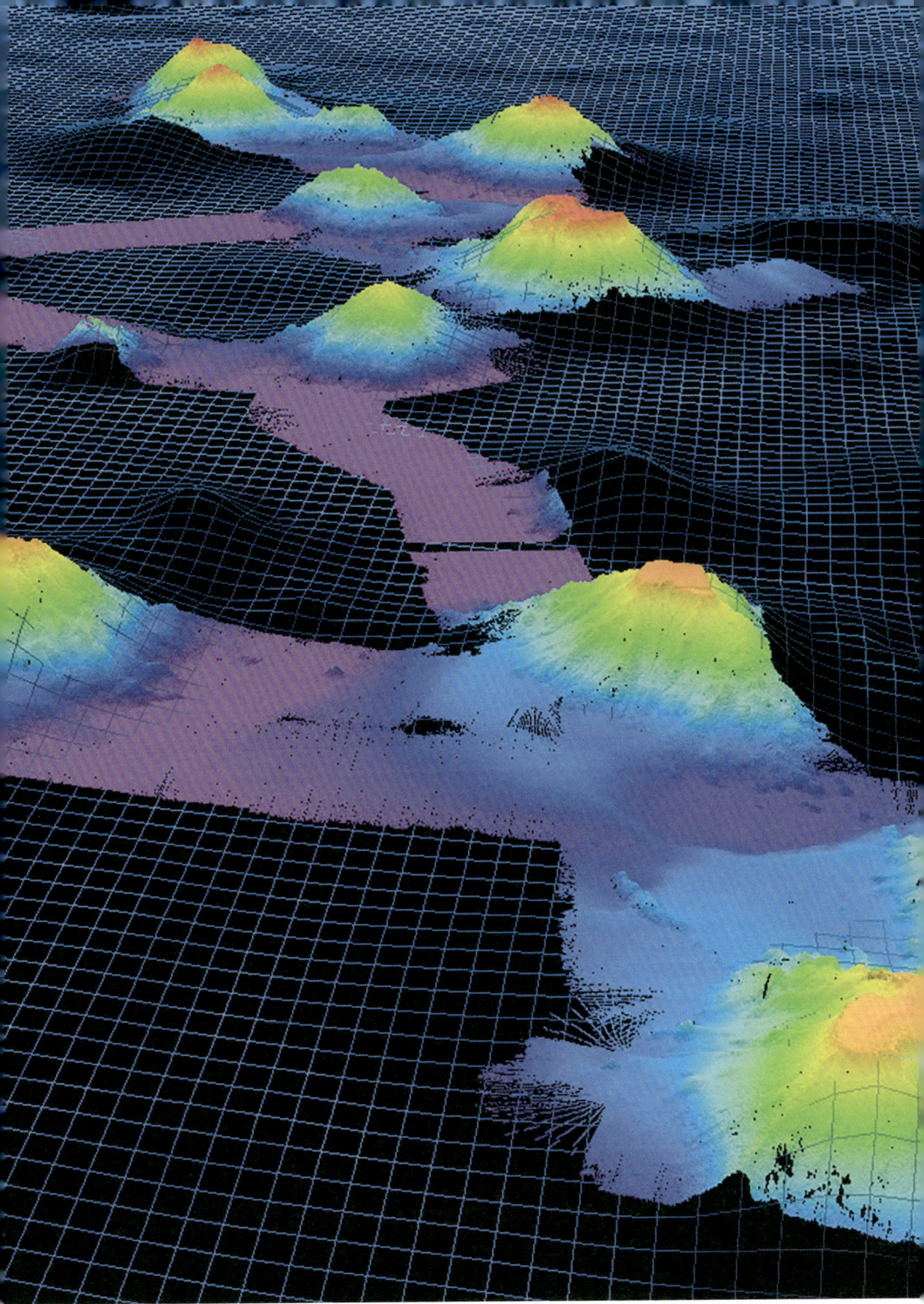

Multibeam sonar is enabling scientists to discover and map seamounts such as these located in the Atlantic Ocean, off the coast of Massachusetts. More seamounts are found each year, thereby enriching our understanding of the importance of these formations to marine life.

A huge thunderstorm builds over a tiny coral island in the Maldives. The country comprises over 1,100 islands which, while seeming a large number, is only a tiny fraction of the more than 600,000 islands in the world.

A green turtle swims in the South Pacific off the coast of French Polynesia. It is the only herbivorous species of turtle and eats mostly seagrass and algae. Females are know to undertake long migrations betweer feeding grounds and the beaches of their birth where they return to lay their own eggs.

beaked whale, travels from higher latitudes to the Azores during autumn. Part of an elusive family – so far we know of twenty-two species – they are rarely encountered and therefore difficult to study, so as yet we do not understand why so many of them make this journey.

Porbeagle sharks, shortfin mako sharks, yellowfin tuna and blue marlin are all much more commonly caught around seamounts than anywhere else in the ocean. It is hard to say for sure if we find them there because that is where we fish for them or because that is where they spend the most time, but it is clear that they regularly travel between seamounts and that seamounts are important places for these and many other pelagic species.

From what we know so far it appears that many ocean travellers, especially whales, dolphins, tuna and sharks, deliberately navigate to seamounts. It is thought they locate them through each seamount's unique effect on ocean currents and its geomagnetic signature. This signature was created when the volcanic rock that formed the seamount cooled and solidified, locking iron atoms in the rock in a particular form. The magnetic field in the seamount tends to fade over time, adding to the uniqueness of the signature.

But seamounts are not only important to visiting ocean travellers; many have an equally precious community of locals. Dense stands of deep-water corals have been found on seamounts all over the world. The first surprise in this discovery was that corals are able to grow at all in conditions so fundamentally different from where we find the better-known, shallow reef-building corals. Shallow corals need warm water with lots of sunlight for their algae to photosynthesise. However, deep-water corals grow in cold water below the depth where photosynthesis is possible. Instead, the polyps inside the coral reach out their tentacles to grab particles of food – typically microscopic creatures and fragments of dead marine organisms known as marine snow. This is a much less efficient way of getting food than reliable daily sunlight, so deep-sea corals are far slower growing than their shallow-water relations, increasing in size by no more than a single millimetre per year and living for hundreds of years.

Fascinatingly, it appears that in parts of the world with dominant seasons the deep-water corals are also affected by the time of year. When it is sunnier and warmer the surface waters are more productive so the volume of marine snow increases and the deep-water coral grows more strongly. In the winter the surface waters have less sunlight so less phytoplankton, and the deep-water coral growth slows accordingly.

They may be lesser known and slower growing, but in their own way deep-water corals are no less spectacular than the more photographed shallow-water coral reefs: thickets of bright white zigzag coral grow next to deep-red soft octocorals with their polyps bursting out of them like blossom, the massive skeletons of blue coral interspersed with sponges, brittle stars and snake stars. Slow-growing, long-lived fish patrol this old-growth coral forest, octopuses hunt among the branches and fans, and spider crabs strut about the ancient structures.

Seamounts are vital habitats for deep-water coral because coral needs a hard base on which to grow. On the muddy floor of the open ocean, a mountain of hard volcanic rock is an excellent landing point for the young free-floating coral known as planulae. When they settle and grow they are sessile, meaning they remain in the same place for ever, forming not reefs but mounds or stands. Depending upon the species of coral, these stands can tower dozens of metres high or spread out, mushroom-like, horizontally. Scientists have discovered seamounts with thriving stands of coral thousands of years old, the oldest so far being black corals dating back to the Bronze Age.

They have also found many which have been reduced to rubble.

In the second half of the twentieth century fishing fleets were commissioned with bigger ships fitted with modern devices, namely on-board freezers and processing capacity to keep the catch fresh until it is landed, radar and GPS to make navigation easier, sonar to help locate underwater features and schools of fish, and stronger nets and winches that allow bigger gear to be deployed to greater depths and over rougher terrain. These distant water fleets were able to travel long distances to fish anywhere in the world, and with the development of global tracking systems and data availability it became possible for initiatives such as Global Fishing Watch to map where they were operating around the world. Overlaying the fishing boats with the known locations of seamounts revealed that, just like whales, sharks and dolphins, the distant water fleets travelled great distances to get to seamounts, their sonar detecting the same massive aggregations of fish scientists had discovered.

The powerful trawlers dragged strong weighted nets over

the seamounts, often hauling up centuries-old coral and sponge gardens along with the fish they were targeting. It was like clear-cutting a forest to catch deer. The heavy nets were too powerful for the slow-growing delicate ecosystems to endure and the distance between seamounts meant it made economic sense to trawl the entire mount, often several times over, rather than move to a new location. The destruction was often absolute.

Seamounts are particularly vulnerable to bottom trawling and as a result these fisheries generally follow a boom-and-bust trajectory. However, it is not just trawling that causes problems. Pelagic long-line fishing over seamounts and around oceanic islands has high levels of by-catch of vulnerable protected species such as sea turtle and shark. This is hardly surprising as those species travel to seamounts regularly and in large numbers, so unless steps are actively taken to avoid it, they will inevitably be caught on the lines and in the nets.

While in practice industrial fishing often takes far too many fish, and uses highly damaging equipment, there is no biological reason why we cannot catch reasonable quantities of fish without destroying the ocean ecosystem; indeed, many fishing communities have been doing exactly that for millennia. There are fish species found in coastal waters and on the continental shelf that spawn regularly and grow fast. If we have the will and foresight to look after them – catching only sustainable numbers with low-impact fishing gear – these fisheries could provide food and livelihoods for ever. But seamounts are quite different.

Damaged seamount ecosystems take hundreds, even

thousands, of years to recover. On a human timescale, it is likely that a trawled seamount will never yield the same volume of catch again. Many of their resident species grow slowly and those that travel to reach seamounts are often both already threatened and of vital importance to the wider ocean.

The vast majority of seamounts discovered to date are in international waters and most of the industrial fishing around them is carried out today by distant-water fleets from just a handful of countries, often heavily subsidised by their governments. Yet the species and habitats they have destroyed were part of a shared global commons, belonging to all of us and the rest of nature. Given the disproportionate importance of seamounts to marine life, and their relatively low economic benefit to fisheries, it seems that one of the clearest win-wins for all life on Earth would be to protect them for the good of us all.

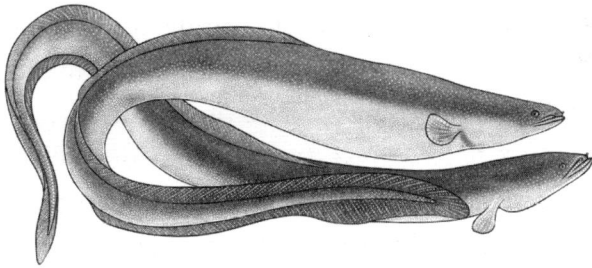

PAPAHĀNAUMOKUĀKEA

Papahānaumokuākea is a deeply sacred place. This is where the Kumulipo, the ancient Hawaiian creation chant, tells us all life forms originate, beginning with a single coral polyp. Native Hawaiian cultural practitioners come here during their lifetimes to connect with the gods and, after death, it is where all spirits return. When the time came to name it, the elders chose Papahānaumokuākea in honour of the earth mother Papahānaumoku and the sky father Wakea, whose union resulted in the creation of the entire archipelago of Hawai'i.

Papahānaumokuākea is one of the most extraordinary ocean areas on the planet. It is a source of abundance. It is āina momona. It is a place where regeneration happens. Rebirth, recovery, renewal. It is a place that when we are looking out at the demise that is happening in so many places on the planet, reminds us of what the planet should be and what the ocean looks like when it is allowed to be able to function in its natural state.

It has been a quarter of a century since Aulani Wilhelm first made the almost 2,000-kilometre journey from Honolulu to Midway in Papahānaumokuākea, and when she speaks of it her voice still carries a respect for this collection of islands, reefs and seamounts, combined with the sense of privilege she felt at being allowed to visit. In the years since then she has served as superintendent for the Papahānaumokuākea Marine National Monument, advised the White House on marine conservation and been one of many indigenous Hawaiians who persuaded successive US presidents to enact a far-reaching plan developed by the local community to turn Papahānaumokuākea into one of the most remarkable and inspirational marine recovery stories on our planet.

Hawai'i can feel geographically isolated, but Papahānaumokuākea is truly remote. Right in the centre of the Pacific, its nearest land masses are Japan to the west, Baja California to the east, Russia to the north and Queensland, Australia, to the south-west. The Pacific is so big that it could comfortably hold all the land on Earth within it. On a map, Papahānaumokuākea itself appears as a tiny spot within the Pacific's immense span, yet stretching top to bottom it still covers a distance equivalent to that of London to Rome. Its remote location and rich waters make this collection of islands and atolls home to an abundance of wildlife and a beacon for migrating species.

But while it has always been remote and sacred, it has not always been insulated from the world. In the nineteenth century its seabirds were killed in their hundreds of thousands for eggs and feathers while their guano was harvested for fertiliser. In 1909 President Roosevelt was so concerned about the seabird slaughter that he put some immediate protections in place in an attempt to stem the destruction. Also during the course of the nineteenth century and into the twentieth century marine species, many of which travelled vast distances to reach Papahānaumokuākea, were killed in astonishing numbers by hunters from distant lands. It was akin to a vast self-service market for animal products. Millions of sharks, whales and birds were butchered. One single ship was recorded killing 1,500 seals in a single short visit, and the population of Hawaiian monk seals was reduced to just 1,000 individuals. Oyster beds were decimated, with over 150,000 black-lipped pearl oysters harvested at just two reefs. Feather hunters killed vast numbers of seabirds. Sea cucumbers, reef fish and sharks were targeted for food and bait, and entrepreneurs from a rabbit cannery released huge numbers of their stock, which promptly ate much of the vegetation on many islands.

But it was the sea turtles that brought the scale of the destruction home to Aulani. 'When I was born, fifty years ago, there were just sixty-seven Hawaiian green sea turtles left. Sixty-seven!' she says. 'Perhaps *the* most iconic Hawaiian marine species, so important to our culture and yet just barely hanging on.'

'The night gives birth to rough-backed turtles,' the Kumulipo creation chant tells us. The *honu*, or Hawaiian green turtle, is as central to Hawaiians' life today as it would have been to the generations of ancestors who sang this chant. Its graceful, calm confidence in the water is reminiscent of that of many Hawaiian people, who seem to have a natural affinity with the ocean.

This population of green sea turtles were nearly all born on the small islands of French Frigate Shoals within Papahānaumokuākea. These tiny spits of sand, rarely more than a metre or so above sea level, are unimaginably small, fragile dots in the immense expanse of the Pacific. Hatching from eggs buried in the sand for almost two months, the juveniles who survive the gauntlet run of beaks and teeth as they cross the sand and swim through the shallows, will spend several years in the open ocean. Out there they feed on floating rafts of vegetation and eventually, when they are large enough, head towards the main Hawaiian Islands to forage. Green turtles are often described as the only herbivorous species of sea turtle living off algae and seagrasses. This is mostly true, but they have also been observed eating discarded fish, invertebrates and even sponges. From this largely plant-based diet they grow to between 90 centimetres and 1.25 metres long and, despite their name, will often be dark brown or olive in colour.

As adults, the females settle into a rhythm for much of the rest of their seventy-year life. They undertake a regular 2,000-kilometre migration from feeding grounds close to the shore of the Hawaiian Islands to their beach of birth in Papahānaumokuākea in order to lay their eggs. It is estimated that 96 per cent of the population of Hawaiian green turtle nest on just a few small islands in Papahānaumokuākea. It is easy, therefore, to see why their global numbers declined so dramatically when the islands were targeted. In 1882 a single ship recorded taking 103 green turtles in just three days from an island in Papahānaumokuākea. Other visitors dug up turtle eggs to eat. Later, as nets became bigger and more durable, many green turtles were caught as by-catch or snagged on drifting, discarded ghost nets. By the 1970s the Hawaiian green turtle was all but extinct.

Like many seamounts and oceanic islands, Papahānau-mokuākea is used by species as a refuge or oasis in a desert of open ocean. But this remoteness was its biggest danger. Being far from any human can often be a benefit to wildlife, but in the case of Papahānaumokuākea the remoteness made it easy for people from many nations to visit and take whatever they wished. It would be easy to claim that because no one lived there then it didn't matter to anyone; a very European way of looking at the world, but nothing was further from the truth.

Papahānaumokuākea's immense significance to Native Hawaiian culture meant Hawaiian fishermen and elders witnessing the damage to fish stocks and the wider habitat were desperate to change things – they just needed a moment for everyone else to pay attention. The impetus came with a mass coral bleaching.

Papahānaumokuākea, or the Northwestern Hawaiian Islands as they were then known, had received some protection on land for nearly 100 years, dating back to Roosevelt's initiative to safeguard seabirds. But so far there had been little action to look after life in the ocean. In political terms, protecting small patches of land is easier than protecting hundreds of thousands of square kilometres of ocean, but biologically that makes little sense because, for small islands like these, the land and the ocean are truly one ecosystem. The seals and seabirds which spend time on the land suffer if the marine food they need is depleted, and protecting a turtle nesting beach is of limited benefit if the breeding females are dying in fishing nets. But in 1998 everything changed.

Aulani remembers clearly the moment when the first major global bleaching event attributed to climate change swept through many of the world's reefs. It was big news and for the first time governments were being asked by their citizens what they were going to do about it.

The US Congress and President Bill Clinton asked the question, 'Where are the coral reefs of the United States?' Most people said Florida, but a few advisers said, 'No. The Hawaiian Islands and specifically the Northwestern Hawaiian Islands.' Well, it wasn't a place that many people in the US administration had heard about. So they wondered what was happening. What was there? What did we need to know?

This desire to understand whether US reefs were being impacted by climate change set off two parallel endeavours that would ultimately come together to spectacular effect. The first was a significant investment of time and money in scientific enquiry and data collection in Papahānaumokuākea. Regular rapid ecological assessments were commissioned involving a combination of research from many different fields simultaneously focused on the same location. The approach brought together many areas of study, including plants, mammals, birds, reptiles, coral, invertebrates, pollution and temperature.

For an ecosystem like Papahānaumokuākea, rapid ecological assessments are an ideal way of getting a picture of how pressures on both land and sea are interacting and understanding where the whole system might be out of balance, perhaps due to overfishing of a single species or from ocean water heatwaves caused by climate change. But all the data in the world means little if there isn't the desire to change things. Too often we know what needs to be done but the will or support is lacking. Not so on this occasion.

271

The attention of the US government gave Native Hawaiian elders and fishers the opportunity they had been waiting for. They had long been concerned about damage to marine life around their sacred islands and had been working on a plan to restore it. The science, local knowledge and political resolve all came together at exactly the right moment.

While it was an opportune moment that ignited the restoration project, Aulani observes, a quarter of a century later, that the reason it has endured is due to something timeless:

Because Native Hawaiians are part of this place and our culture is actually evolved from our land, our sea, our winds, our waters, the stars that rise and set every day above us, our perspective is completely intertwined with its protection.

In December 2000 President Bill Clinton signed the executive order to create the Northwestern Hawaiian Islands Coral Reef Ecosystem Reserve. It was a major moment for the Hawaiian people and for global marine conservation. But it was just the beginning.

Aulani had a front-row seat as the reserve grew and grew. She served as assistant reserve manager from its inception, and then reserve manager until 2006. Then, inspired by Jacques Cousteau's film footage from the reserve, George W. Bush became the third US president to give protection to the Northwestern Hawaiian Islands by declaring it a national monument. It was a clever political move as, while declaring somewhere a sanctuary takes a long time to work its way through legislation, a national monument can be put in place almost immediately by a president and can only be revoked by an act of Congress. Aulani was made superintendent of the monument and a few months later President Bush granted the Native Hawaiians' wish and officially changed the name from Northwestern Hawaiian Islands to Papahānaumokuākea.

Over the next decade Papahānaumokuākea thrived. New species were discovered, known species recovered and its cultural significance shone brightly as a powerful reminder of intergenerational responsibility for looking after our world. When President Barack Obama visited in 2016, he expanded the area of the Papahānaumokuākea National Monument fourfold, right out to the limits of US territory. Now as large as it could ever be, it became what was then the largest marine protected area on the planet, covering over 1.5 million square kilometres – more than all the US national parks combined.

While the headlines are often given to the admittedly impressive roll-call of multiple US presidents, both Democrat and Republican, supporting Papahānaumokuākea, Aulani is quick to give the bulk of the credit for this remarkable achievement to the hundreds of Native Hawaiian elders, scientists, fishers and community leaders who worked tirelessly over decades.

I fully believe that the quarter-century of durability of this marine protected area is because it has the strong, intergenerational support of our Native Hawaiian community. Political support is often ephemeral. But when you have community support, when you have deep cultural ties to not only the place but the management mechanisms, the decision-making that is going into continuing to protect this place, you can have durability.

Thanks to them we no longer have to wonder what impact protecting a large part of the ocean might have – we can actually see for ourselves.

Wisdom has seen it all. No one knows her precise age but she is certainly over seventy years old, meaning that on her yearly visits to Midway Atoll in Papahānaumokuākea she will have witnessed many changes. Wisdom returns to the exact same spot on Midway Atoll where to date she has laid at least thirty-five eggs.

Wisdom is a Laysan albatross – as far as we know the oldest of her kind in the world. Researchers first put a band on her in 1956 and eagerly await her return each year. Aulani describes Laysan albatross or *mōlī* as looking 'a bit like a giant seagull with amazing Cleopatra eyes – one of the most magnificent creatures in the world'. Albatross have long been of great significance to superstitious sailors – the sighting of one bestowing good fortune on a voyage, its absence foretelling an ominous future. It is an appropriate talisman as albatross are the ultimate mariners.

A perfectly proportioned 2-metre wingspan enables the Laysan to ride the winds or use the updraft from waves to soar effortlessly. A bird can cover vast distances with pinpoint accuracy. One was recorded travelling from Washington State on the US mainland back to Midway covering an average of 550 kilometres a day; another found its way back to the tiny Midway Atoll from a spot in the Philippines almost 7,000 kilometres away. They can stay at sea for years. Indeed, for the first three to five years after fledging most will never touch land, even for a moment. They feed by diving to catch fish, squid and crustaceans and avoid dehydration through an ingenious evolutionary adaptation – what birders call a tube-nose. Albatross and their relations, such as petrels and shearwaters, have two nasal tubes on top of their beaks above which specialised glands filter out salt from their bloodstream along with just enough water to create a saline solution; this runs out through the nostrils or 'tubes', enabling the bird to drink seawater without the salt levels in their blood becoming dangerous. These tubes may also help them detect scents such as plankton blooms or large shoals of fish.

The Laysan albatross.

After their long wanderings they return to the same square metre on the same atoll to be reunited with the same mate. Together they brood a single egg. In honour of their astonishing flying and wayfaring skills, *mōlī* have a special place in Native Hawaiian culture. As Aulani explains,

The mōlī were revered. They fly highest and closest to the heavens, so their feathers were used to indicate our highest of chiefs, and thanks to protections mōlī populations have rebounded forty-fold. Almost the entire species of mōlī, the Laysan albatross, exist in Papahānaumokuākea – without it, there would be no mōlī.

Today Papahānaumokuākea has the largest albatross colony on the planet, home to nearly all the world's population of both Laysan and black-footed albatross. It is also the largest seabird rookery on Earth with over 14 million residents each year.

The size of Papahānaumokuākea has been vital to its recovery. Species like shark, seal, green turtle and tuna need big areas. Small marine protected areas also have their place, but for large-scale recovery in species that traverse great distances, you need scale. And this scale has spawned a recovery beyond anything seen on our planet for hundreds of years. Until Papahānaumokuākea National Monument, it was assumed that no protected area could be big enough to help populations of fish species that migrate over long distances, such as yellowfin tuna. But the size of the national monument has meant that the spillover into waters just outside the protected area is so substantial that tuna catch here has increased by over 50 per cent. For the Hawaiian economy and culture, to which tuna is central, this is very good news.

And it has had huge implications for the way the world views marine protection. It shows that creating much larger no-take zones can restore populations of species that spend their lives travelling the open ocean. It is one of the most important results for Aulani, who is keen to use this example to moderate the apparently intractable positions of conservationists and fishers. 'Protecting the ocean doesn't mean being anti-fishing,' she says. 'Fisheries are critical for human well-being – we Hawaiians know that as well as anyone. The goals of healthy fisheries and conservation are the same: more fish, greater abundance, healthier ocean.'

Hawaiian green sea turtle populations have also bounced back. Nowadays it is hard to spend any time swimming, surfing or boating in the waters off Hawai'i without seeing one gliding through the ocean, nibbling on seagrasses or popping its head out of the water to breathe and look around. Astonishingly, they are also regularly seen basking on the beaches of Papahānaumokuākea – all green turtles go ashore to lay eggs but only in Papahānaumokuākea have they also been seen basking; it is one of many wonderful stories from this most remarkable of marine protected areas.

The white, sandy beaches are busy with endangered Hawaiian monk seals, too, while the surrounding waters are home to twenty-four species of whale and dolphin. Enormous tiger sharks travel from thousands of kilometres away to patrol the shallows just metres offshore.

Almost every inch of inland space has been similarly reclaimed by seabirds; indeed, it is now the most important place on our planet for many threatened species. Shearwaters, fairy terns and petrels swirl above in huge clouds, and the larger islands are now home to the world's biggest colonies of albatross. If one of Wisdom's chicks lives for sixty or seventy years, who knows how rich in life Papahānaumokuākea could become.

The recovery here has exceeded all expectations and the future may bring wildlife gatherings the like of which no living human has ever seen. But those who established Papahānaumokuākea are quick to point out that while it can be a refuge for marine species it can't be protected from climate change or pollution. If almost the entire species of Hawaiian green turtle nests on one tiny sand spit just above sea level then how can they be kept safe in the long term? No amount of local conservation can stop rising sea levels.

The story of Papahānaumokuākea is an inspirational one. The question now is how you inspire others to follow its lead. Aulani has a different perspective from the normal arguments in favour of marine protected areas.

People would ask us, 'What are you protecting it from?' And our answer was, 'That's the wrong question. The question is what are we protecting it for?' Papahā-naumokuākea was set aside for its intrinsic value – for its own sake – and for its value to our culture.

Perhaps this is the most important lesson the world can learn from Papahānaumokuākea. While it has had sizeable economic and political benefits, it has fundamentally succeeded because it matters to the Hawaiian people. There is a tendency in conservation these days to focus on the language of economics in order to show a value to the natural world that business and markets respect. Of course this is important, but Papahānaumokuākea suggests that the value to our culture and our sense of self – the value of a place simply for being what it is – is a powerful reason to protect special places in our ocean *and* to ensure that they endure beyond political cycles or market swings, that they are places which are both special to our ancestors and valuable to those that come after us.

8

SOUTHERN OCEAN

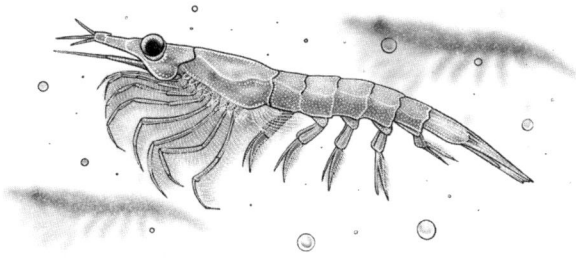

FILMING WITH ELEPHANT SEALS, SOUTH GEORGIA

I first visited South Georgia at the height of the southern summer in December 1980. The male elephant seals had returned from their winter feeding a month or so earlier to stake their claim to the best parts of the beaches and to the harems of females that had formed there. Most of the males were near starving because they had not returned to the ocean for risk of losing their hard-earned places in the social hierarchy. They had been living off nothing except their blubber for perhaps the previous three months, but even so their bulk was still impressive and occasionally intimidating. The wind which can whip across the Southern Ocean for days on end, causing problems for any sound recordist in its vicinity, had died down suffi-ciently for us to think I could walk among them and be filmed and recorded talking to camera.

We found a beach occupied by perhaps thirty or forty elephant seals. I planned to thread my way through them, describing their breeding and social hierarchy. I had a stick to ward off any particularly aggressive males if they resented my approach. It felt a little inadequate.

The most powerful males, the 'beach masters', each gather a harem of females. The beach masters spread out along the shore, making a distinctive low-frequency guttural sound to establish dominance. Up close, the noise vibrates through you, and I had to work hard to stay both far enough away to keep my piece to camera audible while close enough for the beach master we were focused on to stay in shot. Perhaps I got a little too close because he spun towards me faster than his great bulk would suggest possible, clearly ready to defend his territory and the prizes within. It was a reminder of the strength required to remain a master of the beach in the great Southern Ocean.

The seas around South Georgia and the other sub-Antarctic islands are excellent locations for cold water marine life. The islands here are actually the pinnacles of a largely submerged underwater ridge, the Scotia Ridge, which once formed part of a land bridge connecting Antarctica to South America. They also stand in the path of the largest ocean current on Earth – the mighty Antarctic Circumpolar Current. When this immense current, esti-mated to carry 150 times more water than all the rivers on land combined, hits the Scotia Ridge, it forces enormous quantities of nutrient-rich water to the surface, creating vast blooms of phytoplankton and the beginnings of one of the most abundant food chains on Earth.

The southern elephant seals are near the top of this food chain. Up to 5 metres long and weighing over 4 tonnes, they are lumbering and ungainly on land. But in these near freezing waters they are masters of their surroundings. They may spend months at a time at sea, feasting on squid, sharks, rays, shellfish and whatever local shoals of fish they come across. They are prodigious divers and regularly descend as deep as 2,300 metres while holding their breath for twenty minutes or more.

The blubber they develop over winter keeps them warm in the low temperatures and then sustains them through the long months of starvation as they maintain their ownership of a place on the beach throughout the breeding season. But it is that very adaptation that nearly led to their downfall.

When Captain Cook visited South Georgia in 1775, his descriptions of the vast seal colonies fell on receptive ears. Sealers began to target many of the species that were found there to obtain their blubber. For the most part, the elephant seals were left alone, as they were much harder to butcher than fur seals. But before long the whale populations of the Southern Ocean became so depleted that the whalers turned to the elephant seals in order to render their blubber into the oil that their harpoons could no longer deliver.

As the nineteenth century approached its end, a quirk of economics saved the southern elephant seals. Owing to the distances travelled and the manpower required, seal hunting was only profitable if large numbers could be killed. Once the population became low and dispersed, it was no longer worthwhile. So the remnants of the once vast herds were able to cling on until a complete ban on hunting was imposed.

By the time we filmed them, in 1980, their numbers were beginning to recover, and today there are at least 750,000 southern elephant seals. It is a rare and happy moment when I can reflect on a place where I filmed early in my career and know that the scientists and film crews that now visit those same beaches are seeing wonders even more astonishing than those I encountered.

DA

Head past the Roaring Forties, through the Furious Fifties, deep into the Screaming Sixties and you reach a continent so wild and remote it has no indigenous peoples, no permanent settlement, and was not even properly mapped until 1983. The term 'last wilderness' gets attached to a surprising number of places in our ever-more connected world, but if there is one region where it indisputably still applies it is the Southern Ocean and the continent it encircles: Antarctica.

Few, if any, places on Earth are so defined by their latitude. Terms such as 'screaming sixties' are given to the winds at each of the latitudes – Screaming Sixties being the winds at 60 degrees south of the equator and so on – giving character to those concentric circles radiating from the South Pole through the frozen land mass of Antarctica, past a ring of year-round sea ice, a further ring of seasonal sea ice and out into the Southern Ocean itself.

You need to be tough – or at least very well adapted – to survive here. On the ice, male emperor penguins spend four months without food in some of the harshest conditions on Earth, each keeping a single egg warm enough to survive the winter under a special flap of skin known as a brood pouch. On the sea floor sponges, sea spiders, crustaceans and worms all grow far larger than their warm water relatives, though the exact reason for this cold water gigantism is as yet unknown. In water so cold it would freeze the fish of tropical and temperate waters, crocodile icefish have evolved antifreeze proteins in their blood which limit the growth of ice crystals, making the freezing point of their blood slightly below the freezing point of seawater. Arguably that isn't even the most surprising thing about their blood: crocodile icefish have no red blood cells, so their blood is translucent. No one is sure of the evolutionary value of this adaptation; in fact it should be detrimental as without red blood cells it is harder to transport oxygen around their bodies. However, to compensate, they have evolved scale-free skin that allows them to absorb oxygen directly from the seawater.

The Southern Ocean is truly a place of extremes.

It was formed approximately 30 million years ago when tectonic shifts caused the Drake Passage to open up, separating South America and the northern tip of Antarctica and creating the youngest of the world's five ocean basins. There have been many technical and political debates about the exact extent of the Southern Ocean, but, when viewing a satellite image of the region, it makes sense to think of it as extending from the coast of Antarctica to connect the five great capes of the southern hemisphere, namely Africa's Cape of Good Hope, South America's Cape Horn, New Zealand's South West Cape, Tasmania's South East Cape, and Australia's Cape Leeuwin. It is typically between 4,000 and 5,000 metres deep yet its cold surface waters are among the most nutrient-rich in the world and as a result are home to some of the greatest gatherings of marine life ever recorded. Hundreds of fin whales swim so close together their breath appears to form low-level clouds as it condenses above the water; huge rafts of penguins porpoise – jump out of the water – every few metres as they swim at high speed; and on remote islands the beaches thrum with seals and seabirds.

The main reason for this superabundance is the perpetual mixing of water caused by the extreme conditions in this wild region at the bottom of our planet. As sunlight fades each winter and months of darkness descend, the ocean near Antarctica begins to freeze. Tiny ice crystals known as frazils form on the surface. The freezing process takes fresh water out of the sea, leaving the salt behind and creating some of the heaviest, saltiest water anywhere in the ocean. Trillions of tonnes of this dense water sink thousands of metres down into the base of the Southern Ocean. Meanwhile out at sea powerful persistent westerly winds, uninterrupted by any land mass, blow surface waters away and nutrient-rich ones are drawn up from the deep to replace them. This process drives the largest ocean current on our planet, the Antarctic Circumpolar Current. By any measure it is a massive and powerful force. It is the world's only truly global ocean current, encircling the Earth in an uninterrupted irregular loop flowing in a clockwise rotation around Antarctica. It begins south of latitude 40 degrees, in places is 200 kilometres wide and is still exerting its force on the water 4,000 metres beneath the surface.

The Antarctic Circumpolar Current's vital role in mixing the world's oceans is thought to be a major reason why Antarctica remains permanently frozen. The Indian, Atlantic and Pacific oceans all bring warmer water into the Southern Ocean and it is this current that mixes them all up, transporting them in a broadly circular route around the Southern Ocean, ensuring there is a constant barrier of cold water between the ice of Antarctica and warmer surface waters flowing down from the equator.

These unique conditions create perhaps the most important feature for wildlife in the Southern Ocean, the Antarctic Convergence. While its exact location fluctuates seasonally, the Antarctic Convergence is a real biophysical line, similar to the tree line in the Arctic. The Arctic tree line marks the latitude that separates the region where trees can still grow from that further north where it is too cold and only tundra, ice and ocean endure. Similarly, the Antarctic Convergence delineates where the colder, denser waters from the south sink beneath the warmer waters coming from further north. This causes upwellings of nutrients that catalyse vast blooms in phytoplankton which, in turn, fuel the food chains of marine life in the Southern Ocean. So different are the surface water temperatures either side of the convergence that much of the air-breathing marine life either side is distinct and rarely crosses it. True

Antarctic species such as the Weddell seal and emperor penguin can only permanently live in the cold waters south of the Antarctic Convergence.

The same is true for what is arguably the most important creature in the Southern Ocean, the Antarctic krill. If you are in the Southern Ocean and see a reddish hue to the water, whether just a couple of metres wide or stretching for several kilometres, you'll know two things. First, you are south of the convergence, and second, you are looking at tens of thousands of very small animals which are likely to attract dozens of much larger ones. Antarctic krill are small shrimp-like crustaceans with bright red pigmentation in their shells. Like many cold water species, they grow slowly and live for a surprisingly long time. Their eggs hatch in the dark depths of the Southern Ocean and the larvae spend several days swimming to the surface to feed on phytoplankton. It then takes three years for them to grow to their adult length of 5 to 6 centimetres.

Krill have many unique adaptations to allow them to survive the Antarctic winter. As the dark season descends, with less and less light to use in photosynthesis, phytoplankton decline. With this – their main food source – limited, krill slow their metabolism and where possible switch their diet to alternatives such as ice algae, detritus on the seabed and even zooplankton. More remarkably, studies have shown female krill shrinking in winter and regressing their external sexual physiology, thereby effectively returning to their juvenile life stage, when their energy needs were smaller. This downsizing enables them

David Attenborough and an elephant seal, St Andrew's Bay, South Georgia, 1992. David had first filmed in South Georgia in 1980 for *Life on Earth*, and returned over ten years later for a series focusing entirely on Antarctica.

A vast colony of king penguins in St Andrew's Bay, South Georgia.

An iceberg with a large colony of chinstrap penguins in the Southern Ocean near Coronation Island. In the distance are three krill trawlers with

some whale spouts close to the vessels. Krill underpin the Southern Ocean ecosystem and in recent years krill fishing has grown significantly.

A giant petrel provides a sense of scale as it flies in front of this wave-sculpted iceberg floating in the Southern Ocean near Antarctica.

to use their own body protein as fuel to last them through the lean times. With the return of spring they regain their sexual characteristics and become fully mature just in time for the breeding season.

That is, if they aren't eaten first.

Predator–prey relationships in the Southern Ocean are often simpler than in other habitats. Frequently the food chain hinges on krill. Penguins, fish, seals and squid all have krill as a substantial part of their diet. But in the case of baleen whales – filter feeders such as the humpback, blue and sei, named after the baleen plates they have in their mouth to sift the water and trap food – krill are a very big part of a very simple food chain. Blue whales, for example, eat krill in enormous numbers. There can be 10,000 krill in a cubic metre of water and a blue whale can swallow them all in a single gulp. Whales are so dependent upon krill that pregnancy rates of humpback whales were found to correlate with the availability of krill in an area, with higher densities of krill resulting in more calves the following year. As every student of statistics learns, correlation does not mean causation, but in this case the high correlation combined with lack of other possible factors does indeed suggest that a link between krill populations and humpback births is worth taking seriously. So it is fortunate that Antarctic krill are one of the most numerous species in the world, with an estimated 700 trillion of them in the Southern Ocean.

At its simplest, life in the Southern Ocean is founded upon the extreme conditions creating currents and upwellings. These bring nutrients from the deep that fuel phytoplankton growth, which feeds krill that are themselves the food for many other species which live or migrate here.

And we too have migrated here in waves of discovery and exploitation.

For centuries what lay beyond the known land of the southern capes was the subject of debate and mystery. It captivated influential figures such as the early Egyptian astronomer Ptolemy, who believed a spherical Earth required land at its far south to keep it balanced in space, and the sixteenth-century Flemish cartographer Abraham Ortelius who created what is now considered to be the first modern world atlas, the evocatively titled *Theatre of the World* (*Theatrum Orbis Terrarum*). This beautiful map includes an annotation for 'unknown southern land' between the South American archipelago Tierra del Fuego and the bottom of the world.

Over the next 200 years ships sporadically ventured further into the cartographic abyss of the Southern Ocean – sometimes on purpose but often as a result of hostile sailing conditions causing them to veer off course. Those that wound up there reported ice floes, rocky islands and vast numbers of penguins and seals. Convinced by these reports that the fabled southern land must exist, in 1771 the British charged James Cook with finding and claiming it.

It was a time of true exploration and contrasts. The London that James Cook lived in already had Buckingham Palace, St Paul's Cathedral and Oxford Street, yet no one who lived in this thriving capital city on the verge of the industrial revolution was aware that an entire continent on our planet even existed. Cook never reached the land mass of Antarctica, but he was the first to discover many of the defining features of the Southern Ocean. Many of his on-board livestock died from the sudden drop in temperature when he crossed what we now know to be the Antarctic Convergence. He proved that the Southern Ocean flowed without interruption the whole way round the world, and discovered conditions so hostile to human life as to dispel the hopes of many Europeans of claiming Antarctica as a settlement or resupply location for international trade and exploration. But perhaps what would have the most long-term impact on the wildlife of the Southern Ocean was Cook's report of huge numbers of seals and whales. The age-old travel companion of 'discovery' is 'exploitation', and it was soon to set sail.

In 1786, not long after Cook had sighted fur seals and whales around the sub-Antarctic islands of South Georgia and South Sandwich, the British and American hunters arrived. In the thirty-five years that followed it has been estimated that more than 1 million fur seals were taken from South Georgia alone. And it didn't stop there. From that day in 1786 until the moratorium on commercial whaling some 200 years later, millions of fur seals, Weddell seals and elephant seals were slaughtered in the Southern Ocean. Of the estimated 3 million whales killed by people worldwide during the twentieth century, approximately half were from the Antarctic region. The whalers would frequently head towards places where krill swarms were usually found, as often the whales would be there too. While it is likely they knew the whales were hunting the krill, they couldn't have known that the upwellings were essentially a map to find baleen whales – find the upwellings and you find the nutrients, the krill and the whales.

Word soon got around about the rich pickings to be had from this ocean. Sealers and whalers braved the treacherous conditions, and by 1822 there are records of many essentially collapsed seal colonies, with only a few scattered individuals escaping capture as the effort of hunting them in small numbers was no longer worth the reward.

It can be tempting to look back and demonise sealing and whaling, but context is everything. It was undeniably brutal, the conditions for workers were appalling, and there are many examples of the exploitation of poor communities by the companies and governments behind the trade. But there is a big distinction between the early hunters killing for materials they knew no other way to obtain – principally oil – and the industrial whaling and sealing of the late twentieth century when alternatives for most uses of whale parts existed.

The impact of removing millions of top predators from the Southern Ocean is uncertain because, while inventories of the whaling and sealing were made for commercial purposes, no one had cause to study the wider ecosystem. It could be reasonably assumed that the slaughter of marine mammals might result in a superabundance of krill and fish as their main predators were gone, but studies of whales and seals currently returning to the Southern Ocean suggest otherwise. We now know that high populations of whales in the water actually stimulate phytoplankton blooms. Whale faeces contains high concentrations of iron, nitrogen and phosphorus, and in general whales defecate at or near the surface. Their faeces fertilise the water and feed the growth of microscopic plants. Researchers from Stanford University have described the whales as 'mobile krill processing plants'. Krill are rich in iron, so by eating vast quantities of krill and then defecating, the whales concentrate and release iron that is otherwise locked away in the bodies of the krill and make it available for phytoplankton to use. As a result, it is now thought that more whales actually lead to more krill, not less.

Bans on both commercial sealing and whaling in the mid-twentieth century have been extraordinarily effective. All seal and whale species in the Southern Ocean are recovering and many are doing so much faster than anyone predicted. There are today an estimated 3.5 million fur seals and 750,000 elephant seals as well as increased numbers of all species of baleen whale living in or visiting the Southern Ocean. We must note, however, that some whale species are much slower to recover than others: for example, blue whales are recovering, but there are still only between 750 and 2,000 Antarctic blue whales, so there is a long way to go before they reach their natural population. But the general trend is one of recovery, and the wild, rich Southern Ocean waters have responded with krill numbers also bouncing back.

Our recent history in the Southern Ocean is not only one of exploitation. It has also become a celebrated stage for discovery, science and global cooperation. When we think of exploration and discovery in this most extreme of environments, it is the heroic feats of endurance against the odds of Amundsen, Mawson, Scott and Shackleton that perhaps most commonly spring to mind. Yet the twentieth century also saw a quieter, but no less impressive, era of expeditions.

Today there are some 5,000 people in summer and 1,000 in winter working on research expeditions in Antarctica and the Southern Ocean. They are all living in a part of the world owned by no one and set aside for peaceful scientific enquiry. Science bodies from the fifty-seven countries to date who have signed the Antarctic Treaty organise the expeditions and all share their data with each other. It is one of the greatest examples of nations putting aside differences in the interests of science, and for this reason Antarctica is often described as the world's laboratory. The expeditions have discovered many new species, provided invaluable monitoring of the ozone hole and been central to climate modelling and prediction. It is easy to take all this for granted but it very nearly didn't happen, and there's no guarantee it will continue.

On 1 December 1959 twelve countries – Argentina, Australia, Belgium, Chile, France, Japan, New Zealand, Norway, South Africa, the Soviet Union, the United Kingdom and the United States of America – all put their territorial claims on ice and signed the Antarctic Treaty to regulate international relations with respect to the only continent without an indigenous human population. Against the backdrop of the Cold War, this was an extraordinary achievement.

But while the land of Antarctica was set aside for science,

the seas surrounding it remained a free-for-all. Fishing for krill began in the early 1960s with the Soviet Union proposing, incorrectly as we now know, that a lack of whales had led to an abundance of krill which could be used in a myriad of products, including butter made from krill. But many scientists disagreed and the fishing of krill became part of a growing concern for the future of life in Antarctic waters. This concern led to a treaty called the Convention for the Conservation of Antarctic Marine Living Resources and an ongoing international commission that still exists today known as CCAMLR (Commission for the Conservation of Antarctic Marine Living Resources).

CCAMLR extended the ambition of the Antarctic Treaty all the way to the Antarctic Convergence, thus bringing species such as krill into the scope of member nations' activities in that region. The hope was that decades of wise stewardship of the Antarctic continent might now extend to the Southern Ocean. But laudable as the intent and ambition of CCAMLR is, the practical realities of this wild and remote region cause significant problems. The tale of the Patagonian toothfish is perhaps the embodiment of this.

Like many Southern Ocean species, Patagonian toothfish are long-lived and slow-growing. They take in excess of ten years to reach maturity and are typically found between 200 and 2,000 metres beneath the wild waves hunting for squid, crustaceans and finfish. The customers who eat them as steamed 'Chilean seabass' fillets are, in all likelihood, probably not picturing a 2-metre-long, 100-kilogram deep-sea fish with a very long snout, two rows of teeth and a large mouth being dragged from the depths of the Antarctic waters.

The prized Chilean seabass was in such demand that it gained the nickname 'white gold'. Meanwhile the fishing boats, pursuing it with lines several miles long, became infamous for the albatross and other seabirds they caught in pursuit of this prized fish. CCAMLR attempted to regulate the catch, but the area was so vast, and the demand so high, that initially the illegal and unreported catch was almost certainly many times higher than the agreed limit. The approach to fishing short-lived, fast-breeding species of the mid-latitude ocean was being applied to the slow-growing, long-lived species of the far Southern Ocean. Disaster loomed. There was simply no biological way for species like toothfish to keep pace with our demands when it took so long for them to even grow to breeding age. If we carried on this way, the stock of toothfish would be in serious danger.

While it is likely there are still significant unregulated and unreported toothfish catches, recent years have seen bold attempts to protect the Southern Ocean, and CCAMLR has had some success in reducing the seabird by-catch from toothfish fisheries and regulating fisheries within its jurisdiction. In 2016 the largest marine protected area on Earth was established in the Ross Sea, covering a massive 1.55 million square kilometres (approximately 1.1 million of which is fully protected) and home, at different times of year, to more than 50 per cent of the South Pacific Weddell seals, a quarter of all emperor penguins and a third of the world's Adélie penguins.

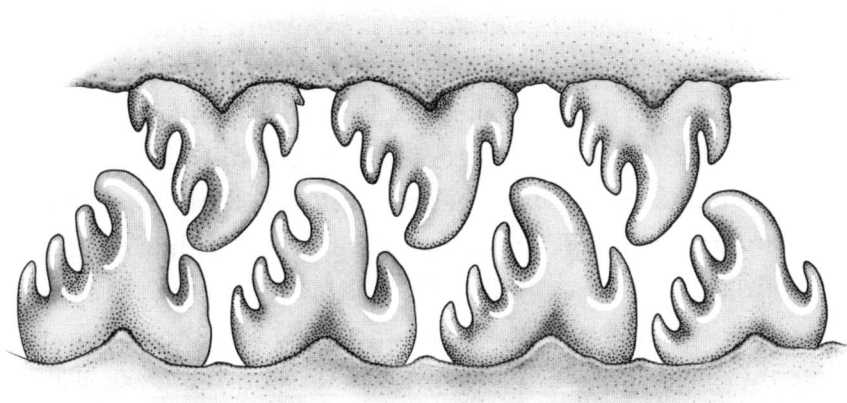

The highly evolved teeth of the crab-eater seal,
native to the Southern Ocean. The shape allows them
to sieve the water and filter out tiny crustaceans
such as krill. They rarely, if ever, eat crabs.

Buoyed by this achievement, scientists and conservation groups have together proposed three huge new marine protected areas in East Antarctica, the Weddell Sea and the Antarctic Peninsula. If agreed, they would together make up nearly 4 million square kilometres of the Southern Ocean. Proposals like this are recognition that this region is the true definition of global commons, owned by no one yet vital to all of us.

Still, as with much of our marine world, the future for the Southern Ocean is unclear.

On the one hand the Antarctic Treaty, now stretching out through the work of CCAMLR to encompass much of life in the Southern Ocean, is one of the most forward-thinking international agreements humanity has so far come up with. The desire to preserve an area beyond national boundaries for peaceful exploration and science in order to benefit all humanity is truly laudable. But at the same time the challenges are immense.

Today enormous boats from faraway countries such as Norway and China drag millions of krill out of the Southern Ocean in massive nets. The ships have on-board processing facilities which turn the krill into pet food, feed for fish farms and health supplements. So sophisticated are these factory boats that they can remain at sea for a year at a time. There is active debate over whether to reduce the current level of legal krill catch to allow for the needs of marine species such as whale and penguin. But regardless of percentage volume, surely it is also worth questioning whether manufacture as pet food and health supplements is the best use of a species which underpins one of the most important food chains on Earth.

And that is before we consider the role of climate change.

The British Antarctic Survey describes the Southern Ocean as the world's largest heat and carbon sink. Their studies reveal that almost three-quarters of human-induced warming to date has been absorbed via the Southern Ocean. It is an area vital to our future, but we are only just beginning to understand it. The science so far suggests that as the Southern Ocean becomes warmer, more old ice will melt and less new ice will form. As a result, its surface waters will tend to be less dense so less water will sink to the deep, reducing the ocean's ability to absorb carbon. Simultaneously, as the ocean cycle slows, less of the ancient deep water will be brought to the surface, which will consequently lack the nutrients required to fuel the growth of phytoplankton, which would draw down carbon from the atmosphere. This is a classic example of a negative feedback loop – and in this case one with global consequences. The Southern Ocean is like an engine room for global ocean currents. As its ocean cycle slows it will in turn slow the conveyor belt of ocean water travelling around the world, reducing its ability to transfer heat from the waters of the tropics to colder regions. This will have major destabilising effects around the world, including the slowing down of the Gulf Stream.

Many of the first warning signs of our fast-heating world are felt at the extreme north and south. But, as polar scientists now caution, what happens at the poles does not stay at the poles.

For now, though, the Southern Ocean is absorbing more and more carbon dioxide from the atmosphere. This is, of course, helpful for climate change in the short term, but it is also making its waters more acidic. Calcium carbonate dissolves in acidic water, which is bad news for creatures such as shellfish which use calcium carbonate to construct their shells and skeletons. Over time the ocean's acidification could cause food chains to slow or even collapse. A 2021 study further found that rising acidity in the ocean could obstruct development of Antarctic krill at the embryonic stage, which would be concerning for much of life in the Southern Ocean.

Antarctic scientists readily acknowledge that we are only just beginning to understand this complex environment. If one thing is clear, it is that an intact Southern Ocean is our ally. Perhaps not to those who face its 20-metre-high waves, temperatures of minus 50 degrees Celsius and winds of over 50 knots. But for those of us who live in calmer climes there are many reasons to be grateful to the Southern Ocean and the brave individuals who cooperate to understand the one continent on Earth where none of us live but which all of us need.

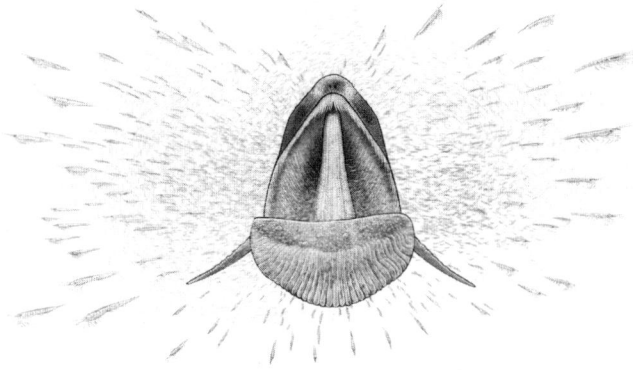

SOUTH GEORGIA AND SOUTH SANDWICH ISLANDS MARINE PROTECTED AREA

Inigo Everson looked out from King Edward Point, not far from the research station that would bear his name on one of its buildings sixty years later to honour decades of scientific work on South Georgia. 'I didn't see a single fur seal,' he remembers. It was the mid-1960s and in those days each sighting of a fur seal would be diligently recorded in the log books. Today, with 3.5 million fur seals in the sub-Antarctic islands, a British Antarctic Survey (BAS) scientist would not dream of noting a single individual sighting.

Fur seals had been hunted on South Georgia since at least 1786 and were all but exterminated by the 1830s. There was a minor recovery in population, but this had led to a recommencement of sealing in the 1870s, and soon they were almost wiped out again. Although by 1960 fur seal numbers had recovered a little from the low point of approximately 200 individuals, the population was still tiny and fragmented. Full-scale recovery had yet to take hold. The very idea of them reaching 3.5 million less than sixty years later would have seemed remarkable to any BAS scientist.

Now in his early eighties and living in south-east England, Inigo had first travelled to South Georgia in 1964 as a recent graduate on board the RRS *Shackleton*. He had enjoyed searching in rock pools as a young boy, read many books on the ocean (including Alister Hardy's *The Open Sea*) and had his interest in fisheries stirred by a particular documentary on the herring industry which he remembers to this day. Together, these influences focused his ambition on a career in marine biology. Discouraged by his head-master because of the lack of jobs in the field, he nonetheless pushed on and with support from one of his science teachers chose to study zoology and oceanography at Bangor University.

It was the ideal setting for a young man with a love of the sea. He spent weekends scuba diving, swimming and rummaging in rock pools in the numerous sheltered bays around Anglesey, all the while cementing the central role of the ocean in his identity. Summer vacation work at the Fisheries Laboratory in Lowestoft furthered his education and brought him into contact with a fellow scientist, Nev Jones; Nev had recently returned from the far south brimming with tales of Antarctica. Inigo now knew what he wanted to do. He was accepted to 'study fish' on the remote Signy Island in the South Orkneys. It was the beginning of a life centred around the great Southern Ocean.

Having spent his entire twenty-two years in the UK, Inigo boarded the RRS *Shackleton* full of excitement and romance only to find the ship itself a little underwhelming.

So you arrive in Southampton and see this tiny little ship and think, crikey, what's this thing doing? You go along that jetty – you walk along there – and there must have been about fifteen or twenty feet from the water level to the dock level and you think, this ship is tiny, where am I going? Where am I going to put everything?

The ship turned out to be as robust as its namesake, and four weeks later Inigo arrived on the sub-Antarctic island. He immediately realised that this was the life he had been seeking. He was in awe of the rugged landscape, wildlife he had previously seen only in books, and the people he met, such as Japanese whalers working at Grytviken and Leith.

Looking back now, though, he reflects on how depleted some of the marine life was – whales and fur seals were rarely seen. He set up a regime for catching fish for his year-round study and supplemented his work with occasional trips to visit penguin colonies, skuas, seals and other wildlife.

When two Soviet fishing vessels stopped by, in 1965, Inigo had his eyes suddenly opened to the commercial fishing of krill and finfish in the Southern Ocean. By 1970 the effects of those fisheries had become clear: he attempted to catch fish among the kelp beds for his studies, a task which had been easy in 1967, and drew a blank. The Soviets had taken over 500,000 tonnes; what had once been a common fish was now extremely rare.

The research Inigo and others were doing in the 1960s and 1970s proved vital to both understanding what lived in the Southern Ocean and establishing the damage overfishing could do. If anything, this research was even more pioneering than equivalent work in other parts of the ocean. The fact that there has never been an indigenous human population on these islands means there are no oral records or art, which could have given some longer-term context to how life in and around these waters had been before Europeans arrived. The closest were records from explorers such as Captain Cook, or from sealers and whalers. Tallies of hunting meant that decades later scientists knew there had once been many more seals on the sub-Antarctic islands, but the wider picture of changes to the marine environment and food chains was only beginning to be uncovered. With modern commercial fishing now added to the mix, the need to first understand, and then manage the wildlife of the Southern Ocean was paramount.

The decades of science and international cooperation

that followed would yield spectacular results. When Inigo revisited South Georgia for the first time in many years, during the early 2000s as part of a research survey with fellow BAS scientist Mark Belchier, he was staggered by the sheer abundance of whales. Mark recalls:

Inigo obviously knew the science and intellectually under-stood that the data showed a huge recovery in whales, fur seals and many other species. But it is one thing academically knowing something and quite another thing to be there and see with your own eyes a region you've known so well for so many years just full of whales.

Mark has returned to South Georgia frequently since his trip with Inigo, and testifies that the recovery didn't stop there. Far from it. The baseline still had a way to rise. He says:

There are many, many more now. If you go down there in the summer the place is awash with humpback whales, fin whales, southern right whales . . . even occasional blue whales. Most are in much larger numbers than Inigo would have experienced in the sixties and seventies or witnessed on his research trip in the early 2000s.

What led to this remarkable turnaround? And how much further could it go?

Two big international agreements had put the brakes on the decline: first, the ban on sealing in the mid-1960s, and then the global moratorium on commercial whaling in 1986 which all but stopped the killing of the baleen whales that spend summer in the sub-Antarctic waters feasting on krill. Both were rightly celebrated but in each case numbers of many species of whale and seal were so low that early recovery was slow. Long-lived, slow-breeding species can take a long time to recover, but a corner had been turned and each year the numbers went up. Now it was time to look at the rest of the ecosystem. Fortunately there was a powerful platform to start from: the Antarctic Treaty. It was then, and is still now, one of the most remarkable examples of vision and pragmatism ever committed to international agreement. The British Antarctic Survey captured it rather well:

There are few places on Earth where there has never been war, where the environment is fully protected, and where scientific research has priority. The whole of the Antarctic continent is like this. A land which the Antarctic Treaty parties call a natural reserve, devoted to peace and science.

The year 1980 was the height of the Cold War, a time when politics, suspicion and fear dominated international relations. Yet despite this global context, Inigo along with many scientists and diplomats from the Soviet Union, the United States, the European Union and twenty-four other nations agreed to cooperate over the management, protection and exploitation of these waters. And they decided to do it all by consensus decision-making. The resulting Convention for the Conservation of Antarctic Marine

Living Resources rarely gets the recognition it deserves, but, for anyone who remembers the Cold War period, simply reading the list of countries involved and imagining getting them in a room together, much less agreeing a resource-sharing treaty, reminds you that it was a truly remarkable achievement. Over the subsequent four decades CCAMLR, the international commission charged with implementing the convention, has provided the impetus, forum and structure for both solid scientific work and significant management improvements across the region – in particular by taking an approach to managing the whole ecosystem rather than just single species. For example, in 2008 they prohibited the use of bottom trawling gear in the entire high seas area of the CCAMLR jurisdiction.

Whereas CCAMLR's work covers all Antarctic waters and inevitably juggles a lot of different, and sometimes competing, interests, the government of South Georgia and South Sandwich Islands has the independence and freedom to go further within their own waters. And in doing so they have precipitated one of the most remarkable marine recoveries anywhere on the planet. The South Georgia and South Sandwich Islands Marine Protected Area was established in 2012 and covers an area of just over 1 million square kilometres. For context, this is over five times the size of the UK. Inside this area there is a complete ban on the mining of minerals and hydrocarbons, trawling of the seabed, and the use of heavy fuel oil by vessels.

About a quarter of this protected area is a complete no-take zone, meaning that absolutely no fishing at all can take place within it. Crucially, it covers the most biologically important and sensitive areas, namely all shallow-water systems as far out as 30 kilometres from shore, all seamounts, and all of the deep-sea trenches. These are the areas crucial for breeding, migration and vegetation growth and, in general, are where we tend to fish the most in seas around the world.

But all those involved are quick to stress that the marine protected area is not anti-fishing; instead, it sets out a way we can both catch fish and restore marine life. Mark Belchier has worked for both the British Antarctic Survey and the South Georgia government during its establishment, and speaks of the safeguards in place:

Fishing within the MPA [marine protected area] is well managed and heavily regulated. There must be independent scientific observers on board collecting data, every boat's location must be transmitted in real time so we always know where they are, fishing seasons are always at a time where there is least competition with wildlife, and all fishing vessels need to use a range of mitigation measures to ensure there's minimal damage to seabirds, seals, whales and other potential by-catch.

But there is also a financial reality:

All the money raised from fisheries licences goes back into the research that underpins the MPA and the patrolling and protection of the waters to make sure illegal or unsustainable fishing doesn't occur. It's a fine line but if we closed the whole area to fishing then very soon we would run out of money even to have a single patrol vessel.

Pragmatism like Mark's appears to be vital to restoring the ocean, especially in parts of the world where there are few permanent residents who would have had, in theory, the incentive to manage and protect their own waters for the future. Whatever your view on this, the results in the South Georgia and South Sandwich Islands Marine Protected Area are spectacular.

Imagine a fish over half a metre long and weighing up to 10 kilograms. The fin on its back – the dorsal fin – is split into two parts, with between four and seven spines in the front section and over thirty in the rear. It has a wide, slightly downturned mouth and velvety green-black marbling along its body, with bigger patterns on its face and smaller dots towards its tail. This is the marbled rockcod, and it lives in the relatively shallow areas – down to about 350 metres – around the islands and continental shelves of the Southern Ocean. It is one of the least glamorous but most impressive recovery stories of the initiative.

The marbled rockcod was the first fish species to be commercially targeted in these waters and was particularly heavily fished during the 1960s and 1970s by Soviet vessels. South Georgia was the epicentre of their activity, with catches estimated at over 500,000 tonnes in just two years. The stock nearly collapsed. Fishing for marbled rockcod was banned, but numbers were so low that many thought they could never recover. This pessimism was exacerbated by the fact that it takes six years for the rockcod to become sexually mature and they only spawn once per season. Scientists waited for a sign that the fishing had been halted in time. But for two decades there was none. However, with the added security of the MPA, scientists now, for the very first time, have credible, well-documented evidence of the beginning of a recovery. Marbled rockcod are slowly but surely coming back.

One of the key characteristics of the sub-Antarctic islands is that they are islands. This sounds facetious, but the point is that what happens on the land there affects what happens in the sea. Indeed, many of the species native to South Georgia are amphibious and rely on one realm as much as the other. And, in parallel to the marine protected area, a transformation has also been occurring on land.

The world's biggest project to successfully eradicate a dangerous invasive species took place on South Georgia. For over 250 years rats and mice, which originally arrived on whaling ships, had decimated the wildlife of this fragile island ecosystem. Thousands of years without land-based predators meant that generations of birds had safely laid their eggs on the ground or in burrows. These were easy pickings for the invasive rodents; their population exploded and the birds declined in parallel. This continued for two and a half centuries. Then in 2018, after a massive conservation effort to remove the mice and rats, South Georgia was declared rodent-free. Within just two years the sound of the South Georgia pipit could be heard once again. A small sparrow-sized streaky brown bird, and Antarctica's only songbird, it had sought refuge in small numbers on tiny islands just off the coast of South Georgia. But when the rodents were eradicated the pipit returned to its former range.

Today with its beaches full of seals, pods of whales surfacing offshore and songbirds once again filling the cool air with their warm calls, against the backdrop of ancient mountains and the collapsing remains of whaling stations, South Georgia feels like paradise reclaimed.

But what does the future hold for this remote and precious outpost? Life within the marine protected area should continue to recover, but where it will settle is yet to be determined. Marbled rockcod and whale numbers may continue to rise, but it is possible that in time the increase in seals will start to affect the rockcod population. Or perhaps the recovery of whales will take more of the available food from the species that seals eat, leading in turn to a rebalancing of seal numbers. Only time will tell, but the beauty of this vast marine protected area, with its vital no-take zones, is that it gives nature both time and space to recover a more natural equilibrium – whatever that turns out to be.

The question for many scientists is, of course, how climate change will affect the future of the marine protected area. Extra nutrients pouring into the ocean from fast-retreating glaciers on South Georgia may be artificially stimulating the food-chain recovery at present, but greater storms, more intense rainfall and run-off combined with warmer waters will probably further alter the balance and stability of the ecosystem. As the area warms, slightly more temperate species like king and gentoo penguins will probably fare better in the short term while slightly more Antarctic penguins like macaroni and chinstrap will suffer.

And what about the krill? This tiny crustacean that underpins the entire food web of the sub-Antarctic islands, and the Southern Ocean more broadly, is the source of much controversy. The experts who manage the South Georgia and South Sandwich Islands Marine Protected Area feel they have the balance right within the area itself, allowing a small amount of krill fishing, but only within the winter months when most whales have migrated away. However, outside the area, many observers remain concerned that krill fisheries could compete locally for the same food as whales and that with climate change and ocean acidification the krill stock itself is in need of further protection. The one thing that is clear is that the success of the protected area around South Georgia and South Sandwich boosts the resilience of the ecosystem and the hope of scientists. Whatever the future effects of climate change or human activity in the wider Southern Ocean, this area will now fare better than it would have done before the designation.

Perhaps, though, this is just the start of the great Southern Ocean recovery. Perhaps CCAMLR can once again provide the stage for its members to put aside today's political differences and expand the model of marine protected areas pioneered by South Georgia and South Sandwich to create a network of restoration and sanctuary all the way across the Southern Ocean. And perhaps others could follow South Georgia's lead and act within their own waters, in the hope that in time the benefits will persuade those slower to respond. What we do know is that wherever there is well-managed marine protection, it is likely that life in the Southern Ocean will rebound.

The sub-Antarctic islands may seem far removed from the modern world where most of us live. But many of their most iconic species, from whale to albatross, are only partial residents and seasonally depart the sanctuary of this protected area, travelling thousands of miles through open waters with little or no protection. Today, on their journeys they will be exposed to fishing gear, pollution and boat strikes. They will find former breeding grounds destroyed and food sources depleted. Yet perhaps in time we can all learn from the story of their islands. We can now see the impact of centuries of exploitation, the targeting of single species, the near collapse of food chains. And we can also appreciate the benefits of continuous dedicated scientific study, evidence-based protection and giving our ocean time and space to rebound. We can see the overwhelming recovery and want more of it.

These are all lessons that could radiate far from these remote islands of the wild Southern Ocean.

PART THREE

IN A SINGLE HUMAN GENERATION

TODAY

To truly see the sea, sometimes we must look to the land.

Off the shore, Cape gannets are jostling, swooping and diving, their movements so frantic and chaotic it is difficult to follow the path of any one individual among the melee. The large black-and-white birds are flying far too fast and erratically for their distinctive golden-yellow crowns to be made out, but their size coupled with the fact that they are pursuing a fishing trawler off the coast of southern Africa reveals their identity.

The seabirds crowd around the trawler scavenging for scraps of hake hauled up from the seabed far below. At first glance it could be a scene of abundance, the heaving nets sustaining both humans and birds. But on land the nests reveal a different story.

On a few precious islands off the coast of South Africa and Namibia, Cape gannets construct nests out of mud and guano on the gently sloping sands. Resembling tiny volcanoes with collapsed cones, thousands of these nests stretch down towards the high-tide line, but an increasing number lie empty and some that are occupied shelter malnourished chicks awaiting a parent's return or the contorted carcasses of those that could wait no longer. Cape gannets have nested on these islands for millennia. Indeed, historically, the fishing boats followed the gannets to locate the shoals of fish. But the roles have tragically reversed.

Attracted by the vast quantity of sardines, both gannet and human fishers thrived for generations. But today the sardine fishery has dwindled. Instead, the boats now trawl deeper water for hake while the gannets, unable to reach such depths, scrabble for leftovers. The fishing waste the parents eat when they cannot find enough of their natural food has half the energy of sardines, leaving them in poor condition, and consequently many chicks grow slowly or die. It is unclear exactly how to apportion blame between overfishing and climate change for the loss of the sardines, but the reality is clear enough: here, life in the ocean can no longer support life on land in the way it once did.

Across the world, wherever land meets sea, there are lost generations. Where ancestors were once drawn to rich coastal habitats those today now struggle among the fragmented echoes of a seemingly lost world. In these coastal settlements there are both those who remember how it once was and those who will live to see what our ocean becomes.

An elderly fisherman, up before dawn as old habits die hard, runs his hand absent-mindedly along the rough rock of a sea wall, his fingertips tracing the edges of mooring cleats as his eyes scan the shingle beach he knows better than anywhere else on earth. Save for the calls and cackles of seabirds, the dawn is calm. He remembers a time when this hour would thrum with activity and noise, with boats being readied while the night fishers were returning and unloading – sharing jokes, arguments breaking out. Life. He would have been able to grab a hot drink and breakfast, and everyone would have joked with his son, asking when *he* would be taking over his old man's boat.

But the joke has become hollow. The signs had been there for years, of course: the foreign trawlers, the falling catch, the ever-smaller fish. But the change was creeping, gradual, imperceptible day to day. For the most part life carried on as it always had. Until it didn't. The next generation had little incentive to work so hard for so little. The boats stopped being repaired. The mornings became quiet.

Thirty years have passed – a generation has been refreshed. The fisherman is now the elder, his son a man. The children he sees playing on the beach nowadays have mostly come from the cities for a holiday or to visit relatives who were too optimistic, or too stubborn, to move away.

But change is in the air: a quickening. This is not a state that can endure. His long experience of the sea tells him that by the time these children reach adulthood, everything will have changed once again.

TOMORROW

The time span of a human generation, some twenty-five to thirty years, is a handy benchmark: long enough for real change to happen but not so far into the future as to be meaningless to those alive now.

Young children playing on a beach today will live through perhaps the most consequential time for the human species in the last 10,000 years. By the time they reach middle age they will have witnessed whether we stall climate change or allow it to engulf us; whether we restore the natural world or fundamentally destabilise it; whether the ocean remains our ally or becomes our foe.

As we have seen already, the ocean can recover. Mangroves and kelp forests *can* regrow, whales *can* return and dying coastal communities *can* flourish once again. Such is the potential for recovery that it is entirely possible there could be but a single lost generation. *We* could be the outliers, the ones who unknowingly took too much and protected too little. Just as those that came before us lived alongside an ocean abounding with life and vitality, today's children could grow up witnessing its resurgence.

There may be much to fear in the near future, yet it could also be the most exciting time to be alive. For we now understand how to fix many of the biggest problems we face as a species, and we have centuries of progress to draw on for inspiration. Indeed, in the past hundred years alone we have dramatically reduced infant mortality, suppressed many of our most feared diseases, increased access to education and health care, acquired scientific knowledge that has transformed our understanding of the world, and cooperated on global issues to a degree never seen before. There is, of course, much we still need to improve in our world, but there is good reason to believe that we have the *ability* to improve it. The question is whether we also have the will and the foresight to do so.

So, what could the next thirty years mean for our ocean? There is an inherent problem in answering this question: anyone who has lived long enough knows the folly of predicting the future. Yet, at the same time, if we aren't able to imagine our future ocean then we will surely struggle to aim for it. What we *can* rationally do is take the discoveries and understanding we have gained and use them to plot a plausible route to the restoration of the ocean while at the same time monitoring and adjusting course as new evidence emerges.

The first thing to recognise is that many of us are far removed from ocean life. If we choose to buy seafood at all we tend to do so from supermarkets or restaurants, with little connection to the process of how it got there. However, for tens of millions of people around the world, fishing is their livelihood, a major part of their local community and economy, and the catch is a vital source of protein. While it may feel morally straightforward to demand change from governments or from the big businesses who profit from the ocean, there are many coastal communities for whom the ability to obtain food from the sea is not a choice but an essential way of life. We should therefore take care to ensure that the actions we employ to restore the ocean benefit those communities, not add to the pressures they already face. There is good reason to believe this is possible.

Decades of scientific study and some wonderful examples of marine protection have shown us that life in the ocean can recover – often much faster than life on land – *if* we give it the time and space to do so. In 2022, at the

United Nations Biodiversity Conference in Montreal, nearly 200 countries agreed to protect 30 per cent of the ocean by 2030. Marine protection can allow wildlife to rebound while maintaining or even increasing fish catch, so this target, known as '30x30', *could* be a game-changer for the ocean. However, we must remain clear-sighted about the effort required to deliver it.

Thirty per cent of the ocean's surface is roughly 120 million square kilometres. By comparison, the celebrated Hawaiian marine protected area, Papahānaumokuākea – massive for such a body – is a mere 1.5 million square kilometres, so 30 per cent of the ocean is a very large area indeed.

While scale is important in the ocean, marine protection is not all about size. Many current marine protected areas are not located in the most biologically important parts of the ocean, as one might presume, but instead in the areas where there is least opposition. Furthermore, most are not enforced, and a shockingly large proportion still allow the most damaging fishing methods, such as bottom trawling, within their borders. Even though, on paper, many reserves may appear to be protected, the benefit to marine life can often be questionable.

The good news, once again, is that changing the status quo is a question of will and finance rather than hoping for new inventions or future technologies that may never arrive. We have reliable evidence that marine protected areas can succeed. Granted they may not all work in exactly the same way, over exactly the same time frame and to exactly the same degree; but they work. Implementing

30x30 can be done with the technology we have today and, if we pick the most appropriate areas to protect and monitor them well, we could reasonably expect great improvements in our ocean within a relatively short period.

But the ocean is a dynamic, connected system, and this raises complexities. For example, no matter how well you protect 30 per cent, if you seriously abuse the other 70 per cent then your plan is unlikely to be as effective as you intended. As a result, many scientists instead suggest we commit to protecting 30 per cent of the ocean and sustainably managing the other 70 per cent. But what does fishing sustainably mean?

Under the standard definition, 'sustainability' means 'to provide for the needs of those alive today without compromising the needs of the generations to follow'. Some claim that quite a lot of current fisheries are 'sustainable' because they aim for what is known as 'maximum sustainable yield' – essentially, the most you can fish in one year while still expecting to catch the same amount the following year – yet this causes much disagreement. It has been estimated that when viewed in this way as much as two-thirds of fish stocks are fished sustainably – it appears we are catching a similar amount of many fish stocks each year. However there are counter-arguments: fish stocks at 'maximum sustainable yield' today are about half of what they would have been pre-fishing; fishing data is incomplete and largely based on what is caught rather than what is actually *in* the ocean; and the reported data doesn't adequately allow for factors such as by-catch, ghost nets or the long-term impact of damage to fragile ecosystems through methods such as bottom trawling.

An entire book could be written on this debate. Doing so would stray into valuable but complex issues, such as whether a fish (or fish stock) should exist for its own intrinsic value or simply for its value to us as a resource. For our purposes the key questions are what amount of fishing, by which means, and in which locations will allow both our ocean to be healthy and those communities that rely on seafood to have a decent quality of life.

There are no simple, universal answers, but we *do* know the parameters. We know that some methods of fishing, especially bottom trawling and dredging, are especially damaging. We also know that the ocean will be more resilient to the impacts of climate change if marine life is rich, diverse and plentiful, *and* that marine life is healthier when habitats have the right balance of species – this is the case whether that habitat is mangrove forest, coral reef, kelp or seagrass meadow. All the evidence suggests that by protecting vast areas across our ocean, banning the most damaging fishing methods and investing in detailed independent monitoring of fish stocks, we should be able to dramatically restore ocean life.

But, in a world of such complexity, can the solution really be as simple as creating proper marine protected areas across a third of the ocean and stopping the use of the most damaging fishing approaches everywhere? Yes and no. Even if nothing else were to change as of today, that should achieve a dramatic improvement. But there are wildcards.

Climate change is already with us. Warming water and ocean heatwaves are already melting sea ice, disrupting ocean currents, bleaching coral reefs and causing marine species to migrate. In general, it seems that climate migration is occurring far faster in the ocean than on land – probably because there are fewer barriers to movement. Indeed, many ocean species are already moving further north or south away from the equator and towards colder water. For example, the North Atlantic right whale migration routes off the coast of Canada have shifted further north in recent years, possibly in response to warming waters and changes in prey availability. These new routes bring the whales into areas of more intense shipping traffic, and this creates a fresh challenge to those working to protect this endangered species. It is just one example of what is likely to become an ocean-wide rearrangement. With significant further warming already certain – our emissions to date are yet to have their full influence on the global climate system – locations considered ideal for marine protection today may be different from those that will be looked on as ideal in a further decade or two. We will need to continually monitor the situation and be adaptable in our approach.

Furthermore, there is a reciprocal relationship between tackling climate change and the health of our ocean. The faster we reduce our emissions, the less damage climate change will cause to marine life: the healthier our ocean becomes, the more it will aid us in slowing climate change. As such, we can benefit from a win-win. But to do so we will need to address both sides of this equation at a speed and scale far beyond our efforts to date.

In the last hundred years the human species has gone beyond being merely 'widespread' or 'dominant' to becoming the single biggest force of change on Earth. So profound is our impact on the biosphere that when viewed over geological time spans the damage we have caused is equivalent in magnitude to that of the five previous major mass-extinction events in Earth's history – including the one that ultimately wiped out most of the dinosaurs. Yet for the wealthy and fortunate among us, who are the cause of the bulk of the problem, the damage is easy to ignore. We enjoy good lives. We have reliable food and water. Outside our homes there are birds and trees. In extreme weather we can retreat indoors to relative comfort. As we go about our privileged daily lives the world does not look apocalyptic; the incentive for change can seem far away.

But the old frames of reference no longer apply. For much of our species' history, our impact was local and the threats we faced were either right in front of our eyes or far beyond our influence – they were 'acts of god'. But while we humans are physically much the same as we were 100,000 years ago, in every other sense we are unrecognisable. Today we are a truly global, planet-shaping species whose biggest threats arise from a biosphere that *we* ourselves have destabilised. The poorest feel these threats acutely while many of us in rich nations are only now beginning to realise that relative wealth cannot insulate us indefinitely. To address these new forms of challenge, to stand a real chance of restoring our ocean and stabilising our world, we do not simply need changes in policy, we need to change our perspective. We need to see ourselves as we truly are.

If we can reset our perspective and recognise that we are a species that, despite wondrous inventions, remains entirely dependent upon a stable biosphere. *If* we can recognise that we are a species that is a part of nature, not somehow removed from the natural world by dint of our great intelligence. *If* we can acknowledge that we now have the capacity to alter and damage the biosphere; *then* we can also recognise that the only way out of this situation is to cooperate and solve our shared problems as a truly global species.

Global agreements might be slow to achieve, hard to enforce and full of compromise, but they are now fundamental to large-scale change. Fortunately we have evolved some helpful attributes. Our unparalleled communication skills and ability to share science, evidence and best practice can both work in our favour. One need only look at the Antarctic Treaty or the moratorium on whaling to know that meaningful, world-changing agreements *are* possible. Is it too much, therefore, to hope that we can also make 30x30 or the High Seas Treaty work? Is it beyond us to agree to ban deep-sea mining? To reduce ocean pollution? To share and enforce best practice in aquaculture?

Changing perspective so that we recognise our planet-wide impact does not mean that smaller, local action is no longer relevant – quite the opposite. Not only can seemingly one-off local examples provide the blueprint and evidence for bigger initiatives; individual stories will help change how we see ourselves as a species.

Humans have always been storytellers; ever since the first human civilisations it has been this that has shaped our culture, forged our identity and defined what we view as important. Our shared stories have formed religions and nations; they underpin politics, create our sense of self and the tribes we gravitate towards. It is hard to think of any great social change in modern history – from civil rights to same-sex relationships – that hasn't been shaped or accelerated by individual stories of injustice or hope. And the same can be true of our transition from a species that exploits and despoils the ocean to one that restores and stewards it.

To date we have done such a good job of telling the stories of demise and collapse that many of us can all too easily picture a future ocean of bleached reefs, turtles choking on plastic, sewage plumes, jellyfish swarms and ghost towns where fishing villages were once full of life. Going forward, our stories of innovation, hope and heroes are of equal importance, as they will both show us who we are and inspire what we still have time to become.

They are not tales of some far-off future but of a reality that is within reach for many alive right now. Imagine a couple of children playing on a beach today, perhaps piling up stones, running in and out of the surf or feeding crumbs to baitfish with their grandparents. *They* will grow up to see how this story ends, to see how *our* choices play out. If we use our great discoveries, apply our unique minds and direct our unparalleled communication and problem-solving skills to restoring our ocean, then they will bring their own children into a world where the biggest challenges our species has ever faced have already been navigated. They will witness decades of recovery and restoration. They will see shoals of fish, roosts of seabirds and pods of whales beyond anything anyone alive today has ever laid eyes upon. They will experience the rebirth of coastal communities and the turning point in the stabilisation of our climate. But more than that, they will live in a world where our species, the most intelligent ever to exist on Earth, has moved beyond trying to rule the waves and instead has learnt to thrive alongside the greatest wilderness of all.

AFTERWORD

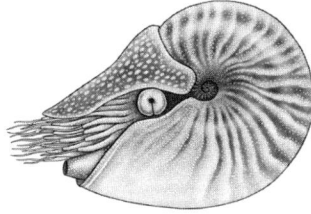

One of the most enjoyable phrases to write in this book was 'remains unknown'. Words to that effect appear many times throughout these pages and each time they hold the tantalising promise of new discoveries to come, new species to find, and a future full of ideas and revelations.

Discovery is one of the great joys of exploring the natural world – whether that's actively seeking or uncovering something you have never seen before or that wonderful moment when a new piece of the puzzle falls into place and you suddenly realise why an aspect of nature functions in the way it does. The last hundred years of humanity's study of the ocean has brought plenty of both. Indeed, even during the writing of this book scientists have revealed incredible new discoveries about life in our ocean.

But with understanding comes responsibility. We now know that our ocean is badly damaged but do we know the future ocean we want? It can be hard to imagine what that future ocean might be because no one alive today has seen a truly rich, thriving ocean. It is normal for each of us to think of the natural world that we saw growing up as 'natural' and only measure improvement or loss against that baseline. Yet, of course, that cannot be true. The ocean that I dived in my youth, some thirty years ago, was both spectacular *and* far diminished from the ocean which David experienced sixty years ago, and so it continues.

But there is no reason for that trend to endure. While there are many more wonders to uncover, we now finally understand all that we need to know to restore our ocean. We have the knowledge but more importantly we have the proof.

We have real-world evidence to show us that we can aim far higher than simply slowing the ocean's decline or stabilising the conditions that we have today, because the most important discovery of recent years is the ocean's capacity to regenerate. Pioneering communities have protected patches of ocean across the world, in all conditions and habitats, and wherever they have done so life has rebounded with spectacular results.

Too often saving the ocean is portrayed as a battle between conservation and fishing, yet we have seen across the world that healthy seas are to the benefit of both fishing communities and wildlife. A revived ocean would provide food, livelihoods, carbon storage, planetary stability . . . and perhaps a few amazing sequences for documentaries as well. We hope that by sharing these tales of discovery, wonder and restoration we may help draw attention to the remarkable gifts that the ocean provides for us all.

Colin Butfield
February 2025

ACKNOWLEDGEMENTS

It may be poetic to imagine a writer (or writers on this occasion) ensconced in some remote location drafting the story they wish to share, agonising over every sentence until it is finally ready to deliver fully formed to the printers. Thankfully, in this case the process was quite different. The reality is that this book emerged from hundreds of conversations, relied on dozens of experts and has been made possible by patient friends and family.

The idea for *Ocean* came about in the margins while making a feature film about the ocean released in 2025. While this book covers quite different stories and subject matter from that film, it is born of a similar hope – to inspire audiences to understand, love and ultimately value our ocean. We are grateful to the creative drive of Toby Nowlan and Keith Scholey and the talents of Doug Anderson, Sam Cooke, Philippa Edwards, Olly Scholey, Kate Streather, Toby Strong, Gavin Thurston and the teams at Open Planet Studios and Silverback Films whose wonderful work on that film kept *us* inspired during the writing of this book.

It is one thing to think of an idea for a book and chat about it for a few months, but quite another to be given the support and encouragement to begin the writing process in

ACKNOWLEDGEMENTS

earnest. For this we are indebted to Robert Kirby and Michael Ridley for believing in this project and partnering us once again with the team at John Murray. Nick Davies, Kate Hewson and Caroline Westmore have been with us throughout, calmly guiding us and bringing in equally expert colleagues including Sarah Arratoon, Helena Dorée, Rebecca Folland, Rosie Gailer, Alice Graham, Hilary Hammond, Louise Henderson-Clark, Alice Herbert, Sophie Jackson, Amanda Jones, Zoë King, Sara Marafini, Janette Revill, Charlotte Robathan, David Roper, Tim Ryder, Megan Schaffer, Ellie Wheeldon, Graham Wild and Corinna Zifko. Thank you all for the passion and care you have put into this book.

They also helped us find collaborators whose talent has greatly enriched this publication, in particular Yeachin Tsai and her beautiful cover art, our gifted illustrator Jennifer Smith and the many photographers and cinematographers whose hard-won images we are grateful to be allowed to use.

Above all, the tales within these pages are of real individuals, communities and expeditions – true ocean heroes. Thank you all for taking the time to speak with us and for trusting us to share your stories. We hope you feel we have done them justice. We are equally indebted to the countless friends and colleagues who helped us track down stories and research, kindly opening up their own contact books in response to our requests.

But it is, as always, our families to whom we owe the greatest thanks. Writing can be a frustrating process and living with those writing books must be considerably more so. To Sophia, Connor and Josh, sorry for the lost weekends and thank you for all your support.

PICTURE CREDITS

Section 1

Page 1: © Alex Mustard. Pages 2–3: Olly Scholey © Open Planet Studios/Silverback Films. Page 4: Doug Anderson © Open Planet Studios/Silverback Films.

Section 2

Page 1: © Atlantic Productions, from *Great Barrier Reef with David Attenborough* (photograph by Freddie Claire). Pages 2–3: courtesy of NOAA Office of Ocean Exploration and Research, Hohonu Moana 2016. Page 4 above and below: © NOAA/Ocean Exploration Trust.

Section 3

Pages 1 and 2–3: © Olly Scholey. Page 4: Doug Anderson © Open Planet Studios/Silverback Films.

Section 4

Pages 1 and 4: Doug Anderson © Open Planet Studios/Silverback Films. Pages 2–3: © Alex Mustard.

PICTURE CREDITS

SOURCES AND FURTHER READING

Coral Reefs

Aretz, M., and Nudds, J., 'The Coral Fauna of the Holkerian/Asbian Boundary Stratotype Section (Carboniferous) at Little Asby Scar (Cumbria, England) and Its Implications for the Boundary', *Stratigraphy*, 2 (2005): 167–90

Bollati, E., et al., 'Green Fluorescent Protein-like Pigments Optimise the Internal Light Environment in Symbiotic Reef-building Corals', *eLife*, 11 (2022): e73521

Bourne, D. G., Morrow, K. M., and Webster, N. S., 'Insights into the Coral Microbiome: Underpinning the Health and Resilience of Reef Ecosystems', *Annual Review of Microbiology*, 70 (2016): 317–40

Cooper, T. F., O'Leary, R. A., and Lough, J. M., 'Growth of Western Australian Corals in the Anthropocene', *Science*, 335, no. 6068 (2012): 593–6

Cornwall, C. E., et al., 'Resistance of Corals and Coralline Algae to Ocean Acidification: Physiological Control of Calcification Under Natural pH Variability', *Proceedings of the Royal Society B*, 285, no. 1884 (2018): 20181168

CTI-CFF Regional Secretariat, *Regional Plan of Action 2.0: Coral Triangle Initiative on Coral Reefs, Fisheries and Food Security 2021–2030*, Manado, Indonesia: Coral Triangle Initiative on Coral Reefs, Fisheries and Food Security, 2022

Darwin, Charles R., *The Structure and Distribution of Coral Reefs. Being the first part of the geology of the voyage of the Beagle, under the command of Capt. Fitzroy, R.N. during the years 1832 to 1836,* 1st edition, London: Smith, Elder, 1842

Doney, S. C., et al., 'Ocean Acidification: The Other CO_2 Problem', *Annual Review of Marine Science*, 1 (2009): 169–92

Elise, S., et al., 'An Optimised Passive Acoustic Sampling Scheme to Discriminate Among Coral Reefs' Ecological States', *Ecological Indicators*, 107 (2019): 105627

Eyal, G., et al., 'Spectral Diversity and Regulation of Coral Fluorescence in a Mesophotic Reef Habitat in the Red Sea', *PLOS ONE*, 10 (2015): e0128697

Eyre, B. D., et al., 'Coral Reefs Will Transition to Net Dissolving Before End of Century', *Science*, 359, no. 6378 (2018): 908–11

Frieler, K., et al., 'Limiting Global Warming to 2 °C Is Unlikely to Save Most Coral Reefs', *Nature Climate Change*, 3 (2013): 165–70

Goffredo, S., et al., 'Colony and Polyp Biometry and Size Structure in the Orange Coral *Astroides calycularis* (Scleractinia: Dendrophylliidae)', *Marine Biology Research*, 7, no. 3 (2011): 272–80

Gruber, N., et al., 'The Oceanic Sink for Anthropogenic CO_2 from 1994 to 2007', *Science*, 363, no. 6432 (2019): 1193–9

Harborne, A. R., et al., 'Multiple Stressors and the Functioning of Coral Reefs', *Annual Review of Marine Science*, 9 (2017): 445–68

Hawkins, J. P., and Roberts, C. M., 'Effects of Artisanal Fishing on Caribbean Coral Reefs', *Conservation Biology*, 18, no. 1 (2004): 215–26

Hoegh-Guldberg, O., and Bruno, J. F., 'The Impact of Climate Change on the World's Marine Ecosystems', *Science*, 328, no. 5985 (2010): 1523–8

Hughes, T. P., et al., 'Coral Reefs in the Anthropocene', *Nature*, 546 (2017): 82–90

——, 'Spatial and Temporal Patterns of Mass Bleaching of Corals in the Anthropocene', *Science*, 359, no. 6371 (2018): 80–3

Intergovernmental Panel on Climate Change, *Climate Change 2021 – The Physical Science Basis: Working Group I Contribution to the Sixth Assessment Report of the Intergovernmental Panel on Climate Change*, 1st edition, Cambridge: Cambridge University Press, 2023

Lapid, E., Wielgus, J., and Chadwick-Furman, N., 'Sweeper Tentacles of the Brain Coral *Platygyra daedalea*: Induced Development and Effects on Competitors', *Marine Ecology Progress Series*, 282 (2004): 161–71

Lin, T.-H., et al., 'Exploring Coral Reef Biodiversity via Underwater Soundscapes', *Biological Conservation*, 253 (2021): 108901

McClanahan, T. R., and Shafir, S. H., 'Causes and Consequences of Sea Urchin Abundance and Diversity in Kenyan Coral Reef Lagoons', *Oecologia*, 83 (1990): 362–70

Matsuda, S. B., et al., 'Temperature-Mediated Acquisition of Rare Heterologous Symbionts Promotes Survival of Coral Larvae under Ocean Warming', *Global Change Biology*, 28 (2022): 2006–25

Mourier, J., et al., 'Extreme Inverted Trophic Pyramid of Reef Sharks Supported by Spawning Groupers', *Current Biology*, 26, no. 15 (2016): 2011–16

Mumby, P. J., et al., 'Marine Reserves, Fisheries Ban, and 20 Years of Positive Change in a Coral Reef Ecosystem', *Conservation Biology*, 35, no. 5 (2021): 1473–83

Roberts, C. M., 'Effects of Fishing on the Ecosystem Structure of Coral Reefs', *Conservation Biology*, 9, no. 5 (1995): 988–95

Roopin, M., and Chadwick, N. E., 'Benefits to Host Sea Anemones from Ammonia Contributions of Resident Anemonefish', *Journal of Experimental Marine Biology and Ecology*, 370, nos 1–2 (2009): 27–34

Sabine, C. L., et al., 'The Oceanic Sink for Anthropogenic CO_2', *Science*, 305, no. 5682 (2004): 367–71

Sakai, K., 'Effect of Colony Size, Polyp Size, and Budding Mode on Egg Production in a Colonial Coral', *Biological Bulletin*, 195, no. 3 (1998): 319–25

Sheppard, C., *Coral Reefs: A Natural History*, Princeton: Princeton University Press, 2021

Simpson, S., et al., 'Attraction of Settlement-Stage Coral Reef Fishes to Reef Noise', *Marine Ecology Progress Series*, 276 (2004): 263–8

——, 'Homeward Sound', *Science*, 308, no. 5719 (2005): 221

Smith, E. G., et al., 'Screening by Coral Green Fluorescent Protein

(GFP)-like Chromoproteins Supports a role in Photoprotection of Zooxanthellae', *Coral Reefs*, 32 (2013): 463–74

Souter, D., et al. (eds), *Status of Coral Reefs of the World: 2020*, Global Coral Reef Monitoring Network and International Coral Reef Initiative, 2021, DOI: 10.59387/WOTJ9184

Spalding, M. D., Ravilious, C., and Green, E. P., *World Atlas of Coral Reefs*, Berkeley, CA: University of California Press, 2001

Tebbett, S. B., Connolly, S. R., and Bellwood, D. R., 'Benthic Composition Changes on Coral Reefs at Global Scales', *Nature Ecology and Evolution*, 7 (2023): 71–81

Van Oppen, M. J. H., and Blackall, L. L., 'Coral microbiome Dynamics, Functions and Design in a Changing World', *Nature Reviews Microbiology*, 17, no. 9 (2019): 557–67

Wilson, E. C., 'Permian Corals from the Spring Mountains, Nevada', *Journal of Paleontology*, 65, no. 5 (1991): 727–41

Wisshak, M., et al., 'Ocean Acidification Accelerates Reef Bioerosion', *PLOS ONE*, 7 (2012): e45124

Zapalski, M. K., et al., 'Coralliths of Tabulate Corals from the Devonian of the Holy Cross Mountains (Poland)', *Palaeogeography, Palaeoclimatology, Palaeoecology*, 585 (2022): 110745

The Deep

Alfaro-Lucas, J., et al., 'Trophic Structure and Chemosynthesis Contributions to Heterotrophic Fauna Inhabiting an Abyssal Whale Carcass', *Marine Ecology Progress Series*, 596 (2018): 1–12

Anderson, T. R., and Rice, T., 'Deserts on the Sea Floor: Edward Forbes and His Azoic Hypothesis for a Lifeless Deep Ocean', *Endeavour*, 30, no. 4 (2006): 131–7

Ausubel, J. H., Trew Crist, D., and Waggoner, P. E., *First Census of Marine Life 2010: Highlights of a Decade of Discovery*, Washington, DC: Census of Marine Life International Secretariat, 2010

Bianchi, D., and Mislan, K. A. S., 'Global Patterns of Diel Vertical Migration Times and Velocities from Acoustic Data: Global Patterns of Diel Vertical Migration', *Limnology and Oceanography*, 61 (2016): 353–64

Bolstad, K. S. R., et al., 'In-Situ Observations of an Intact Natural Whale Fall in Palmer Deep, Western Antarctic Peninsula', *Polar Biology*, 46, no. 4 (2023): 123–32

Bouchet, P., et al., 'Marine Biodiversity Discovery: The Metrics of New Species Descriptions', *Frontiers in Marine Science*, 10 (2023): 929989

Boyd, P. W., et al., 'Multi-faceted Particle Pumps Drive Carbon Sequestration in the Ocean', *Nature*, 568 (2019): 327–35

Brierley, A. S., 'Diel Vertical Migration', *Current Biology*, 24 (2014): R1074–R1076

Cailliet, G. M., et al., 'Age Determination and Validation Studies of Marine Fishes: Do Deep-Dwellers Live Longer?', *Experimental Gerontology*, 36, nos 4–6 (2001): 739–64

Corliss, J. B., et al., 'Submarine Thermal Springs on the Galápagos Rift', *Science*, 203, no. 4385 (1979): 1073–83

Etter, W., and Hess, H., 'Reviews and Syntheses: The First Records of Deep-Sea Fauna – a Correction and Discussion', *Biogeosciences*, 12, no. 21 (2015): 6453–62

Foley, N. S., Van Rensburg, T. M., and Armstrong, C. W., 'The Rise and Fall of the Irish Orange Roughy Fishery: An Economic Analysis', *Marine Policy*, 35 (2011): 756–63

Giering, S. L. C., et al., 'Reconciliation of the Carbon Budget in the Ocean's Twilight Zone', *Nature*, 507 (2014): 480–3

Global Ocean Science Report: The Current Status of Ocean Science Around the World, Paris: UNESCO Publishing, 2017

Glover, A. G., and Smith, C. R., 'The Deep-Sea Floor Ecosystem: Current Status and Prospects of Anthropogenic Change by the Year 2025', *Environmental Conservation*, 30, no. 3 (2003): 219–41

Harris, P. T., et al., 'Geomorphology of the Oceans', *Marine Geology*, 352 (2014): 4–24

Hein, J. R., Koschinsky, A., and Kuhn, T., 'Deep-Ocean Polymetallic Nodules as a Resource for Critical Materials', *Nature Reviews Earth & Environment*, 1 (2020): 158–69

Jónasdóttir, S. H., et al., 'Seasonal Copepod Lipid Pump Promotes Carbon Sequestration in the Deep North Atlantic', *Proceedings of the National Academy of Sciences of the United States of America*, 112, no. 39 (2015): 12122–6

Jones, E., *The Challenger Expedition: Exploring the Ocean's Depths*, London: Royal Museums Greenwich, 2022

Kelley, D. S., et al., 'An Off-Axis Hydrothermal Vent Field Near the Mid-Atlantic Ridge at 30° N', *Nature*, 412 (2001): 145–9

Kobayashi, T., Nagai, H., and Kobayashi, K., 'Concentration Profiles of 10Be in Large Manganese Crusts', *Nuclear Instruments and Methods in Physics Research Section B: Beam Interactions with Materials and Atoms*, 172 (2000): 579–82

Levin, L. A., Amon, D. J., and Lily, H., 'Challenges to the Sustainability of Deep-Seabed Mining', *Nature Sustainability*, 3 (2020): 784–94

Mayer, L., et al., 'The Nippon Foundation – GEBCO Seabed 2030 Project: The Quest to See the World's Oceans Completely Mapped by 2030', *Geosciences*, 8 (2018): 63

Paulus, E., 'Shedding Light on Deep-Sea Biodiversity – A Highly Vulnerable Habitat in the Face of Anthropogenic Change', *Frontiers in Marine Science*, 8 (2021): 667048

Pitcher, T. J., et al. (eds), *Seamounts: Ecology, Fisheries and Conservation*, Oxford: Blackwell, 2007

Purser, A., et al., 'Association of Deep-Sea Incirrate Octopods with Manganese Crusts and Nodule Fields in the Pacific Ocean', *Current Biology*, 26 (2016): R1268–R1269

Roark, E. B., et al., 'Extreme Longevity in Proteinaceous Deep-Sea Corals', *Proceedings of the National Academy of Sciences of the United States of America*, 106, no. 13 (2009): 5204–8

Robison, B. H., 'Bioluminescence in the Benthopelagic Holothurian *Enypniastes eximia*', *Journal of the Marine Biological Association of the United Kingdom*, 72, no. 2 (1992): 463–72

Sala, E., et al., 'The Economics of Fishing the High Seas', *Science Advances*, 4, no. 6 (2018): eaat2504

Simon-Lledó, E., et al., 'Biological Effects 26 Years after Simulated Deep-Sea Mining', *Scientific Reports*, 9 (2019): 8040

Smith, C. R., Roman, J., and Nation, J. B., 'A Metapopulation Model for Whale-Fall Specialists: The Largest Whales Are Essential to Prevent Species Extinctions', *Journal of Marine Research*, 77, no. 2 (2019): 283–302

Spiess, F. N., et al., 'East Pacific Rise: Hot Springs and Geophysical Experiments', *Science*, 207, no. 4438 (1980): 1421–33

Tarling, G. A., et al., 'Carbon and Lipid Contents of the Copepod Calanus finmarchicus Entering Diapause in the Fram Strait and Their Contribution to the Boreal and Arctic Lipid Pump', *Frontiers of Marine Science*, 9 (2022): 926462

Thiel, H., et al., 'The Large-scale Environmental Impact Experiment DISCOL – Reflection and Foresight', *Deep Sea Research Part II: Topical Studies in Oceanography*, 48, nos 17–18 (2001): 3869–82

Thomas, K. N., Robison, B. H., and Johnsen, S., 'Two Eyes for Two Purposes: In Situ Evidence for Asymmetric Vision in the Cockeyed squids *Histioteuthis heteropsis* and *Stigmatoteuthis dofleini*', *Philosophical Transactions of the Royal Society B*, 372, no. 1717 (2017): 20160069

Wölfl, A.-C., et al., 'Seafloor Mapping – The Challenge of a Truly Global Ocean Bathymetry', *Frontiers of Marine Science*, 6 (2019): 283

Yin, K., Zhang, D., and Xie, W., 'Experimental Whale Falls in the South China Sea', *Ocean-Land-Atmosphere Research*, 2, no. 2 (2023): 0005

Open Ocean

Block, B. A., et al., 'Tracking Apex Marine Predator Movements in a Dynamic Ocean', *Nature*, 475 (2011): 86–90

Branch, T. A., et al., 'Past and Present Distribution, Densities and Movements of Blue Whales *Balaenoptera musculus* in the Southern Hemisphere and Northern Indian Ocean', *Mammal Review*, 37, no. 2 (2007): 116–75

SOURCES AND FURTHER READING

Carmine, G., et al., 'Who Is the High Seas Fishing Industry?', *One Earth*, 3, no. 6 (2020): 730–8

Carwardine, M., et al., *Handbook of Whales, Dolphins and Porpoises*, London: Bloomsbury, 2020

Chiba, S., et al., 'Human Footprint in the Abyss: 30 Year Records of Deep-Sea Plastic Debris', *Marine Policy*, 96 (2018): 204–12

Clukey, K. E., et al., 'Investigation of Plastic Debris Ingestion by Four Species of Sea Turtles Collected as Bycatch in Pelagic Pacific Longline Fisheries', *Marine Pollution Bulletin*, 120 (2017): 117–25

Davies, T. E., et al., 'Multispecies Tracking Reveals a Major Seabird Hotspot in the North Atlantic', *Conservation Letters*, 14, no. 5 (2021): e12824

——, 'Tracking Data and the Conservation of the High Seas: Opportunities and Challenges', *Journal of Applied Ecology*, 58, no. 12 (2021): 2703–10

Duncan, E., et al., 'A Global Review of Marine Turtle Entanglement in Anthropogenic Debris: A Baseline for Further Action', *Endangered Species Research*, 34 (2017): 431–48

——, 'Microplastic Ingestion Ubiquitous in Marine Turtles', *Global Change Biology*, 25, no. 2 (2019): 744–52

García-Alegre, A., et al., 'Seabed Litter Distribution in the High Seas of the Flemish Pass Area (NW Atlantic)', *Scientia Marina*, 84 (2020): 93

Hong, S., Lee, J., and Lim, S., 'Navigational Threats by Derelict Fishing Gear to Navy Ships in the Korean Seas', *Marine Pollution Bulletin*, 119 (2017): 100–5

Houghton, J. D. R., et al., 'The Role of Infrequent and Extraordinary Deep Dives in Leatherback Turtles (*Dermochelys coriacea*)', *Journal of Experimental Biology*, 211, no. 16 (2008): 2566–75

Jarvis, R. M., and Young, T., 'Pressing Questions for Science, Policy, and Governance in the High Seas', *Environmental Science & Policy*, 139, no. 3 (2023): 177–84

Johnson, C. M., et al., *Protecting Blue Corridors: Challenges and*

Solutions for Migratory Whales Navigating National and International Seas, World Wildlife Fund, 2022

Kühn, S., and Van Franeker, J. A., 'Quantitative Overview of Marine Debris Ingested by Marine Megafauna', *Marine Pollution Bulletin*, 151 (2020): 110858

Laist, D. W., 'Impacts of Marine Debris: Entanglement of Marine Life in Marine Debris Including a Comprehensive List of Species with Entanglement and Ingestion Records', in J. M. Coe and D. B. Rogers (eds), *Marine Debris*, New York: Springer, 1997, pp. 99–139

Lebreton, L., et al., 'Industrialised Fishing Nations Largely Contribute to Floating Plastic Pollution in the North Pacific Subtropical Gyre', *Scientific Reports*, 12 (2022): 12666

Letessier, T. B., et al., 'Remote Reefs and Seamounts are the Last Refuges for Marine Predators Across the Indo-Pacific', *PLOS Biology*, 17 (2019): e3000366

Lohmann, K. J., Lohmann, C. M. F., and Endres, C. S., 'The Sensory Ecology of Ocean Navigation', *Journal of Experimental Biology*, 211, no. 11 (2008): 1719–28

Montecalvo, I., et al., 'Ocean Predators: Squids, Chinese Fleets and the Geopolitics of High Seas Fishing', *Marine Policy*, 152 (2023): 105584

Mouritsen, H., 'Long-Distance Navigation and Magnetoreception in Migratory Animals', *Nature*, 558 (2018): 50–9

Mucientes, G., and Queiroz, N., 'Presence of Plastic Debris and Retained Fishing Hooks in Oceanic Sharks', *Marine Pollution Bulletin*, 143 (2019): 6–11

Nelms, S. E., et al., 'Investigating Microplastic Trophic Transfer in Marine Top Predators', *Environmental Pollution*, 238 (2018): 999–1007

——, 'Plastic Pollution and Marine Megafauna: Recent Advances and Future Directions', in Alice Horton (ed.), *Plastic Pollution in the Global Ocean*, Singapore: World Scientific, 2023, pp. 97–138

Pacoureau, N., et al., 'Half a Century of Global Decline in Oceanic Sharks and Rays', *Nature*, 589 (2021): 567–71

Pérez Rodam, M. A., *A Third Assessment of Global Marine Fisheries Discards*, Rome: Food and Agriculture Organization of the United Nations, 2019

Richardson, K., Hardesty, B. D., and Wilcox, C., 'Estimates of Fishing Gear Loss Rates at a Global Scale: A Literature Review and Meta-analysis', *Fish and Fisheries*, 20, no. 6 (2019): 1218–31

Rocha Jr, R. C., Clapham, P. J., and Ivashchenko, Y., 'Emptying the Oceans: A Summary of Industrial Whaling Catches in the 20th Century', *Marine Fisheries Review*, 76, no. 4 (2015): 37–48

Romeo, T., et al., 'First Evidence of Presence of Plastic Debris in Stomach of Large Pelagic Fish in the Mediterranean Sea', *Marine Pollution Bulletin*, 95, no. 1 (2015): 358–61

Sala, E., et al., 'The Economics of Fishing the High Seas', *Science Advances*, 4, no. 6 (2018): eaat2504

Santos, R. G., et al., 'Debris Ingestion by Juvenile Marine Turtles: An Underestimated Problem', *Marine Pollution Bulletin*, 93, nos 1–2 (2015): 37–43

Senko, J., et al., 'Understanding Individual and Population-Level Effects of Plastic Pollution on Marine Megafauna', *Endangered Species Research*, 43 (2020): 234–52

Spotila, J. R., *Sea Turtles: A Complete Guide to Their Biology, Behavior, and Conservation*, Baltimore, MD: Johns Hopkins University Press, 2004

Sumaila, U. R., et al., 'Winners and Losers in a World Where the High Seas Is Closed to Fishing', *Scientific Reports*, 5 (2015): 8481

Unger, B., et al., 'Large Amounts of Marine Debris Found in Sperm Whales Stranded Along the North Sea Coast in Early 2016', *Marine Pollution Bulletin*, 112, nos 1–2 (2016): 134–41

United Nations General Assembly, Agreement under the United Nations Convention on the Law of the Sea on the Conservation and Sustainable Use of Marine Biological Diversity of Areas Beyond National Jurisdiction, New York, 19 June 2023

Zantis, L. J., et al., 'Assessing Microplastic Exposure of Large Marine Filter-Feeders', *Science of the Total Environment*, 818 (2022): 151815

Kelp Forest

Cerda, O., Hinojosa, I. A., and Thiel, M., 'Nest-Building Behavior by the Amphipod *Peramphithoe femorata* (Krøyer) on the Kelp *Macrocystis pyrifera* (Linnaeus) C. Agardh from Northern-Central Chile', *Biological Bulletin*, 218, no. 3 (2010): 248–58

Clendenning, K. A., 'Organic Productivity in Kelp Area', in W. A. North (ed.), *The Biology of Giant Kelp Beds (Macrocystis) in California*, Lehre, Germany: Verlag Von J. Cramer, 1974, pp. 259–63

Daub, C. D., et al., 'Fucoidan from *Ecklonia maxima* Is a Powerful Inhibitor of the Diabetes-Related Enzyme, α-glucosidase', *International Journal of Biological Macromolecules*, 151, no. 8 (2020): 412–20

Davis, R. W., and Pagano, A. M. (eds), *Ethology and Behavioral Ecology of Sea Otters and Polar Bears*, Cham, Switzerland: Springer, 2021

Duarte, C. M., et al., 'Global Estimates of the Extent and Production of Macroalgal Forests', *Global Ecology and Biogeography*, 31, no. 7 (2022): 1422–39

Feehan, C. J., Filbee-Dexter, K., and Wernberg, T., 'Embrace Kelp Forests in the Coming Decade', *Science*, 373, no. 6557 (2021): 863

Hobday, A., 'Abundance and Dispersal of Drifting Kelp *Macrocystis pyrifera* Rafts in the Southern California Bight', *Marine Ecology Progress Series*, 195 (2000): 101–16

——, 'Persistence and Transport of Fauna on Drifting Kelp (*Macrocystis pyrifera* (L.) C. Agardh) Rafts in the Southern California Bight', *Journal of Experimental Marine Biology and Ecology*, 253, no. 1 (2000): 75–96

Jayathilake, D. R. M., and Costello, M. J., 'A Modelled Global Distribution of the Kelp Biome', *Biological Conservation*, 252 (2020): 108815

——, 'Version 2 of the World Map of Laminarian Kelp Benefits from More Arctic Data and Makes it the Largest Marine Biome', *Biological Conservation*, 257 (2021): 109099

Jorgensen, S. J., et al., 'Killer Whales Redistribute White Shark Foraging Pressure on Seals', *Scientific Reports*, 9 (2019): 6153

Krause-Jensen, D., and Duarte, C. M., 'Substantial Role of Macroalgae in Marine Carbon Sequestration', *Nature Geoscience*, 9 (2016): 737–42

Krumhansl, K. A., et al., 'Global Patterns of Kelp Forest Change over the Past Half-Century', *Proceedings of the National Academy of Sciences of the United States of America*, 113, no. 48 (2016): 13785–90

Li, H., et al., 'Transcriptomic Responses to Darkness and the Survival Strategy of the Kelp *Saccharina latissima* in the Early Polar Night', *Frontiers of Marine Science*, 7 (2020): 592033

Lissner, A. L., and Dorsey, J. H., 'Deep-Water Biological Assemblages of a Hard-Bottom Bank-Ridge Complex of the Southern California Continental Borderland', *Bulletin of the Southern California Academy of Sciences*, 85 (1986): 87–101

Mabate, B., et al., 'Fucoidan Structure and Its Impact on Glucose Metabolism: Implications for Diabetes and Cancer Therapy', *Marine Drugs*, 19, no. 1 (2021): 30

Martínez, E. A., Cárdenas, L., and Pinto, R., 'Recovery and Genetic Diversity of the Intertidal Kelp *Lessonia nigrescens* (Phaeophyceae) 20 Years after El Niño 1982/83', *Journal of Phycology*, 39, no. 3 (2003): 504–8

Perissinotto, R., and McQuaid, C., 'Deep Occurrence of the Giant Kelp *Macrocystis laevis* in the Southern Ocean', *Marine Ecology Progress Series*, 81 (1992): 89–95

Pessarrodona, A., et al., 'Carbon Sequestration and Climate Change Mitigation Using Macroalgae: A State of Knowledge Review', *Biological Reviews*, 98, no. 6 (2023): 1945–71

Qin, Y., 'Alginate Fibres: An Overview of the Production Processes and Applications in Wound Management', *Polymer International*, 57 (2008): 171–80

Rogers-Bennett, L., and Catton, C. A., 'Marine Heat Wave and Multiple Stressors Tip Bull Kelp Forest to Sea Urchin Barrens', *Scientific Reports*, 9 (2019): 15050

Schiel, D. R., and Foster, M. S., *The Biology and Ecology of Giant Kelp Forests*, Oakland, CA: University of California Press, 2015

Schiel, D. R., Steinbeck, J. R., and Foster, M. S., 'Ten Years of Induced Ocean Warming Causes Comprehensive Changes in Marine Benthic Communities', *Ecology*, 85, no. 7 (2004): 1833–9

Smith, S. D. A., 'Kelp Rafts in the Southern Ocean', *Global Ecology and Biogeography*, 11 (2002): 67–9

Spalding, H., Foster, M. S., and Heine, J. N., 'Composition, Distribution, and Abundance of Deep-Water (>30 m) Macroalgae in Central California', *Journal of Phycology*, 39, no. 2 (2003): 273–84

Starko, S., Wilkinson, D. P., and Bringloe, T. T., 'Recent Global Model Underestimates the True Extent of Arctic Kelp Habitat', *Biological Conservation*, 257 (2021): 109082

Towner, A., et al., 'Fear at the Top: Killer Whale Predation Drives White Shark Absence at South Africa's Largest Aggregation Site', *African Journal of Marine Science*, 44, no. 2 (2022): 139–52

Vergés, A., and Campbell, A. H., 'Kelp Forests', *Current Biology*, 30, no. 16 (2020): R919–R920

Wernberg, T., et al., 'Climate-Driven Regime Shift of a Temperate Marine Ecosystem', *Science*, 353, no. 6295 (2016): 169–72

Arctic

Albert, C., et al., 'Seasonal Variation of Mercury Contamination in Arctic Seabirds: A Pan-Arctic Assessment', *Science of the Total Environment*, 750 (2021): 142201

Arctic Monitoring and Assessment Programme (AMAP), *Arctic Pollution Issues: A State of the Arctic Environment Report*, Tromsø, Norway: AMAP, 1997

——, *Assessment 2020: POPs and Chemicals of Emerging Arctic Concern: Influence of Climate Change*, Tromsø, Norway: AMAP, 2021

——, *Assessment 2021: Mercury in the Arctic*, Tromsø, Norway: AMAP, 2021

SOURCES AND FURTHER READING

Boltunov, A., et al., 'Persistent Organic Pollutants in the Pechora Sea Walruses', *Polar Biology*, 42 (2019): 1775–85

Davis, R. W., and Pagano, A. M. (eds), *Ethology and Behavioral Ecology of Sea Otters and Polar Bears*, Cham, Switzerland: Springer, 2021

Dietz, R., et al., 'Current State of Knowledge on Biological Effects from Contaminants on Arctic Wildlife and Fish', *Science of the Total Environment*, 696 (2019): 133792

Higdon, J. W., and Ferguson, S. H., 'Loss of Arctic Sea Ice Causing Punctuated Change in Sightings of Killer Whales (*Orcinus orca*) Over the Past Century', *Ecological Applications*, 19 (2009): 1365–75

Higdon, J. W., Hauser, D. D. W., and Ferguson, S. H., 'Killer Whales (*Orcinus orca*) in the Canadian Arctic: Distribution, Prey Items, Group Sizes, and Seasonality', *Marine Mammal Science*, 28, no. 2 (2012): E93–E109

Jay, C. V., et al., 'Walrus Haul-out and in-Water Activity Levels Relative to Sea Ice Availability in the Chukchi Sea', *Journal of Mammalogy*, 98, no. 2 (2017): 386–96

Kinnard, C., et al., 'Reconstructed Changes in Arctic Sea Ice over the Past 1,450 Years', *Nature*, 479 (2011): 509–12

Kwok, R., 'Arctic Sea Ice Thickness, Volume, and Multiyear Ice Coverage: Losses and Coupled Variability (1958–2018)', *Environmental Research Letters*, 13, no. 10 (2018): 105005

Laidre, K. L., et al., 'Interrelated Ecological Impacts of Climate Change on an Apex Predator', *Ecological Applications*, 30, no. 4 (2020): e02071

Letcher, R. J., et al., 'Exposure and Effects Assessment of Persistent Organohalogen Contaminants in Arctic Wildlife and Fish', *Science of the Total Environment*, 408 (2010): 2995–3043

Lewis, K. M., van Dijken, G. L., and Arrigo, K. R., 'Changes in Phytoplankton Concentration Now Drive Increased Arctic Ocean Primary Production', *Science*, 369, no. 6500 (2020): 198–202

Macdonald, R. W., Harner, T., and Fyfe, J., 'Recent Climate Change in the Arctic and Its Impact on Contaminant Pathways and Interpretation of Temporal Trend Data', *Science of the Total Environment*, 342, nos 1–3 (2005): 5–86

Nansen, F., *Farthest North: Being the Record of a Voyage of Exploration of the Ship 'Fram' 1893–96 and the Fifteenth Months' Sleigh Journey (George Newnes, by Arrangement with Archibald Constable)*, Cambridge: Cambridge University Press, 1898

Notz, D., and Stroeve, J., 'Observed Arctic Sea-Ice Loss Directly Follows Anthropogenic CO_2 Emission', *Science*, 354, no. 6313 (2016): 747–50

Pedro, S., et al., 'Blubber-Depth Distribution and Bioaccumulation of PCBs and Organochlorine Pesticides in Arctic-Invading Killer Whales', *Science of the Total Environment*, 601–602 (2017): 237–46

Posdaljian, N., et al., 'Changes in Sea Ice and Range Expansion of Sperm Whales in the Eclipse Sound Region of Baffin Bay, Canada', *Global Change Biology*, 28, no. 12 (2022): 3860–70

Previdi, M., Smith, K. L., and Polvani, L. M., 'Arctic Amplification of Climate Change: A Review of Underlying Mechanisms', *Environmental Research Letters*, 16, no. 9 (2021): 093003

Rantanen, M., et al., 'The Arctic Has Warmed Nearly Four Times Faster than the Globe since 1979', *Communications Earth & Environment*, 3 (2022): 168

Rose, G. A., 'Capelin (*Mallotus villosus*) Distribution and Climate: A Sea "Canary" for Marine Ecosystem Change', *ICES Journal of Marine Science*, 62, no. 7 (2005): 1524–30

Savinov, V., et al., 'Persistent Organic Pollutants in Ringed Seals from the Russian Arctic', *Science of the Total Environment*, 409, no. 14 (2011): 2734–45

Scheuhammer, A. M., et al., 'Effects of Environmental Methylmercury on the Health of Wild Birds, Mammals, and Fish', *AMBIO: A Journal of the Human Environment*, 36, no. 1 (2007): 12–19

World Wildlife Fund, 'The Tallurutiup Imanga National Marine Conservation Area', *The Circle*, 4 (2020): 14–15

Yurkowski, D. J., et al., 'A Temporal Shift in Trophic Diversity among a Predator Assemblage in a Warming Arctic', *Royal Society Open Science*, 5, no. 10 (2018): 180259

SOURCES AND FURTHER READING

Mangroves

Armitage, A. R., et al., 'The Contribution of Mangrove Expansion to Salt Marsh Loss on the Texas Gulf Coast', *PLOS ONE*, 10 (2015): e0125404

Atwood, T. B., et al., 'Global Patterns in Mangrove Soil Carbon Stocks and Losses', *Nature Climate Change*, 7 (2017): 523–8

Bunting, P., et al., 'Global Mangrove Extent Change 1996–2020: Global Mangrove Watch Version 3.0', *Remote Sensing*, 14, no. 15 (2022): 3657

——, 'Global Mangrove Watch: Updated 2010 Mangrove Forest Extent (v2.5)', *Remote Sensing*, 14, no. 15 (2022): 1034

Donato, D. C., et al., 'Mangroves Among the Most Carbon-Rich Forests in the Tropics', *Nature Geoscience*, 4, no. 5 (2011): 293–7

Elhassan, I., 'Occurrence of the Green Sawfish *Pristis zijsron* in the Sudanese Red Sea with Observations on Reproduction', *Endangered Species Research*, 36 (2018): 41–7

Frias-Torres, S., and Luo, J., 'Using Dual-Frequency Sonar to Detect Juvenile Goliath Grouper *Epinephelus itajara* in Mangrove Habitat', *Endangered Species Research*, 7 (2009): 237–42

Gaos, A. R., et al., 'Living on the Edge: Hawksbill Turtle Nesting and Conservation Along the Eastern Pacific Rim', *Latin American Journal of Aquatic Research*, 45, no. 3 (2017): 572–84

——, 'Shifting the Life-History Paradigm: Discovery of Novel Habitat Use by Hawksbill Turtles', *Biology Letters*, 8, no. 1 (2012): 54–6

Goldberg, L., et al., 'Global Declines in Human-Driven Mangrove Loss', *Global Change Biology*, 26, no. 10 (2020): 5844–55

Hogarth, P. J., *The Biology of Mangroves and Seagrasses*, Oxford: Oxford University Press, 2015

Hollensead, L., et al., 'Assessing Residency Time and Habitat Use of Juvenile Smalltooth Sawfish Using Acoustic Monitoring in a Nursery Habitat', *Endangered Species Research*, 37 (2018): 119–31

Igulu, M. M., et al., 'Mangrove Habitat Use by Juvenile Reef Fish:

Meta-Analysis Reveals that Tidal Regime Matters more than Biogeographic Region', *PLOS ONE*, 9 (2014): e114715

Kanno, S., et al., 'Mangrove Use by Sharks and Rays: A Review', *Marine Ecology Progress Series*, 724 (2023): 167–83

Khan, M. M. H., 'Species Diversity, Relative Abundance and Habitat Use of the Birds in the Sundarbans East Wildlife Sanctuary, Bangladesh', *Forktail*, 21 (2005): 79–86

Kristensen, E., et al., 'Organic Carbon Dynamics in Mangrove Ecosystems: A Review', *Aquatic Botany*, 89, no. 2 (2008): 201–19

Leal, M., and Spalding, M. D., *The State of the World's Mangroves 2022*, Global Mangrove Alliance, 2022

McKee, K. L., Cahoon, D. R., and Feller, I. C., 'Caribbean Mangroves Adjust to Rising Sea Level through Biotic Controls on Change in Soil Elevation', *Global Ecology and Biogeography*, 16, no. 5 (2007): 545–56

McLeod, E., et al., 'A Blueprint for Blue Carbon: Toward an Improved Understanding of the Role of Vegetated Coastal Habitats in Sequestering CO_2', *Frontiers in Ecology and the Environment*, 9 (2011): 552–60

Menéndez, P., et al., 'The Global Flood Protection Benefits of Mangroves', *Scientific Reports*, 10 (2020): 4404

Mumby, P. J., et al., 'Mangroves Enhance the Biomass of Coral Reef Fish Communities in the Caribbean', *Nature*, 427 (2004): 533–6

Nagelkerken, I., et al., 'The Habitat Function of Mangroves for Terrestrial and Marine Fauna: A Review', *Aquatic Botany*, 89, no. 2 (2008): 155–85

Naha, D., et al., 'Ranging, Activity and Habitat Use by Tigers in the Mangrove Forests of the Sundarban', *PLOS ONE*, 11 (2016): e0152119

Polidoro, B. A., et al., 'The Loss of Species: Mangrove Extinction Risk and Geographic Areas of Global Concern', *PLOS ONE*, 5 (2010): e10095

Saha, S., et al., 'Shark Diversity in the Indian Sundarban Biosphere', *Journal of Fish Taxonomy*, 23 (2022): 54–8

Sanderman, J., et al., 'A Global Map of Mangrove Forest Soil Carbon at 30 M Spatial Resolution', *Environmental Research Letters*, 13, no. 5 (2018): 055002

Shah, S., *Pandemic: Tracking Contagions, from Cholera to Ebola and Beyond*, New York: Picador, 2017

Spalding, M. D., Kainuma, M., and Collins, L., *World Atlas of Mangroves*, London: Earthscan, 2010

Spalding, M. D., et al., *Mangroves for Coastal Defence: Guidelines for Coastal Managers & Policy Makers*, Wetlands International and The Nature Conservancy, 2014

Tomlinson, P. B., *The Botany of Mangroves*, 2nd edition, New York: Cambridge University Press, 2016

Wedemeyer-Strombel, K. R., et al., 'Fishers' Ecological Knowledge and Stable Isotope Analysis Reveal Mangrove Estuaries as Key Developmental Habitats for Critically Endangered Sea Turtles', *Frontiers in Conservation Science*, 2 (2021): 796868

Worthington, T., and Spalding, M., *Mangrove Restoration Potential: A Global Map Highlighting a Critical Opportunity*, Apollo – University of Cambridge Repository, 2018, DOI: 10.17863

Oceanic Islands and Seamounts

Afonso, P., et al., 'The Azores: A Mid-Atlantic Hotspot for Marine Megafauna Research and Conservation', *Frontiers of Marine Science*, 6 (2020): 826

Bessudo, S., et al., 'Residency of the Scalloped Hammerhead Shark (*Sphyrna lewini*) at Malpelo Island and Evidence of Migration to Other Islands in the Eastern Tropical Pacific', *Environmental Biology of Fishes*, 91, no. 2 (2011): 165–76

Bo, M., et al., 'Persistence of Pristine Deep-Sea Coral Gardens in the Mediterranean Sea (SW Sardinia)', *PLOS ONE*, 10 (2015): e0119393

Booth, D. T., et al., 'Egg Viability of Green Turtles Nesting on Raine Island, the World's Largest Nesting Aggregation of Green Turtles', *Australian Journal of Zoology*, 69 (2021): 12–17

Carreiro-Silva, M., et al., 'Variability in Growth Rates of Long-Lived

Black Coral Leiopathes sp. from the Azores', *Marine Ecology Progress Series*, 473 (2013): 189–99

Clark, M. R., Tittensor, D., and Rogers, A., *Seamounts, Deep-Sea Corals and Fisheries: Vulnerability of Deep-Sea Corals to Fishing on Seamounts Beyond Areas of National Jurisdiction*, Cambridge: UNEP World Conservation Monitoring Centre, 2006

Davies, T. E., et al., 'Multispecies Tracking Reveals a Major Seabird Hotspot in the North Atlantic', *Conservation Letters*, 14, no. 5 (2021): e12824

Finley, C., 'The Industrialization of Commercial Fishing, 1930–2016', in *Oxford Research Encyclopedia of Environmental Science*, online edition, Oxford University Press, 2016, DOI: 10.1093/acrefore/9780199389414.013.31

Fock, H. O., et al., 'Diel and Habitat-Dependent Resource Utilisation by Deep-Sea Fishes at the Great Meteor Seamount: Niche Overlap and Support for the Sound Scattering Layer Interception Hypothesis', *Marine Ecology Progress Series*, 244 (2002): 219–33

Foley, N. S., van Rensburg, T. M., and Armstrong, C. W., 'The Rise and Fall of the Irish Orange Roughy Fishery: An Economic Analysis', *Marine Policy*, 35, no. 6 (2011): 756–63

Isaacs, J. D., and Schwartzlose, R. A., 'Migrant Sound Scatterers: Interaction with the Sea Floor', *Science*, 150, no. 3705 (1965): 1810–13

Jensen, M. P., et al., 'Environmental Warming and Feminization of One of the Largest Sea Turtle Populations in the World', *Current Biology*, 28, no. 1 (2018): 154–9

Kerry, C. R., Exeter, O. M., and Witt, M. J., 'Monitoring Global Fishing Activity in Proximity to Seamounts Using Automatic Identification Systems', *Fish and Fisheries*, 23, no. 3 (2022): 733–49

Leitner, A. B., Neuheimer, A. B., and Drazen, J. C., 'Evidence for Long-Term Seamount-Induced Chlorophyll Enhancements', *Scientific Reports*, 10 (2020): 12729

Letessier, T. B., et al., 'Remote Reefs and Seamounts Are the Last Refuges for Marine Predators Across the Indo-Pacific', *PLOS Biology*, 17 (2019): e3000366

McCauley, D. J., et al., 'Wealthy Countries Dominate Industrial Fishing', *Science Advances*, 4, no. 6 (2018): eaau2161

Mashayek, A., et al., 'On the Role of Seamounts in Upwelling Deep-Ocean Waters Through Turbulent Mixing', *Proceedings of the National Academy of Sciences of the United States of America*, 121, no. 27 (2024): e2322163121

Morato, T., et al., 'Seamounts Are Hotspots of Pelagic Biodiversity in the Open Ocean', *Proceedings of the National Academy of Sciences of the United States of America*, 107, no. 21 (2010): 9707–11

NOAA Office of National Marine Sanctuaries, *2020 State of Papahānaumokuākea Marine National Monument: Status and Trends 2008–2019*, Silver Spring, MD: US Department of Commerce, National Oceanic and Atmospheric Administration, Office of National Marine Sanctuaries, 2020

Pitcher, T. J., *Seamounts: Ecology, Fisheries & Conservation*, Oxford: Blackwell, 2007

Prouty, N., et al., 'Growth Rate and Age Distribution of Deep-Sea Black Corals in the Gulf of Mexico', *Marine Ecology Progress Series*, 423 (2011): 101–15

Roark, E., et al., 'Radiocarbon-Based Ages and Growth Rates of Hawaiian Deep-Sea Corals', *Marine Ecology Progress Series*, 327 (2006): 1–14

Rowden, A. A., et al., 'Paradigms in Seamount Ecology: Fact, Fiction and Future', *Marine Ecology*, 31, series 1 (2010): 226–41

Schmidt Ocean Institute, 'Four New Seamounts Discovered in the High Seas, 2024', press release, 8 February 2024

Silva, M. A., et al., 'Spatial and Temporal Distribution of Cetaceans in the Mid-Atlantic Waters Around the Azores', *Marine Biology Research*, 10, no. 2 (2014): 123–37

Tsukamoto, K., 'Spawning of Eels Near a Seamount', *Nature*, 439 (2006): 929

SOURCES AND FURTHER READING

Southern Ocean

Atkinson, A., et al., 'A Re-appraisal of the Total Biomass and Annual Production of Antarctic Krill', *Deep Sea Research Part I: Oceanographic Research Papers*, 56, no. 5 (2009): 727–40

Basberg, B., and Headland, R. K., 'The 19th Century Antarctic Sealing Industry: Sources, Data and Economic Significance', NHH Department of Economics, Discussion Paper no. 21 (2008), DOI: 10.2139/ssrn.1553751

Branch, T. A., et al., 'Past and Present Distribution, Densities and Movements of Blue Whales *Balaenoptera musculus* in the Southern Hemisphere and Northern Indian Ocean', *Mammal Review*, 37, no. 2 (2007): 116–75

Busch, B. C., *The War Against the Seals: A History of the North American Seal Fishery*, Montreal: McGill-Queen's University Press, 1987

Cheng, L., et al., 'Past and Future Ocean Warming', *Nature Reviews Earth & Environment*, 3 (2022): 776–94

Clarke, A., and Johnston, N. M., 'Antarctic Marine Benthic Diversity', in R. N. Gibson and R. J. A. Atkinson (eds), *Oceanography and Marine Biology: An Annual Review*, London: Taylor & Francis, 2003, pp. 47–114

Costa, D. P., and Crocker, D. E., 'Marine Mammals of the Southern Ocean', in E. E. Hofmann et al. (eds), *Antarctic Research Series*, Washington, DC: American Geophysical Union, 1996, pp. 287–301

Donohue, K. A., et al., 'Mean Antarctic Circumpolar Current Transport Measured in Drake Passage', *Geophysical Research Letters*, 43 (2016)

Forcada, J., et al., 'Ninety Years of Change, from Commercial Extinction to Recovery, Range Expansion and Decline for Antarctic Fur Seals at South Georgia', *Global Change Biology*, 29, no. 24 (2023): 6867–87

Fraser, C., et al., 'An Ocean Like No Other: The Southern Ocean's

Ecological Richness and Significance for Global Climate', *The Conversation*, 6 December 2020

Frölicher, T. L., et al., 'Dominance of the Southern Ocean in Anthropogenic Carbon and Heat Uptake in CMIP5 Models', *Journal of Climate*, 28, no. 2 (2015): 862–86

Government of New Zealand, 'Ross Sea Region Marine Protected Area', press release, 26 October 2016, https: //www.mfat.govt.nz/en/envi ronment/antarctica-and-the-southern-ocean/ross-sea-region-marine-protected-area (accessed 25 May 2024)

Hamabe, K., Matsuoka, K., and Kitakado, T., 'Estimation of Abundance and Population Dynamics of the Antarctic Blue Whale in the Antarctic Ocean South of 60°S, from 70°E to 170°W', *Marine Mammal Science*, 39 (2023): 671–87

Hindell, M. A., et al., 'Circumpolar Habitat Use in the Southern Elephant Seal: Implications for Foraging Success and Population Trajectories', *Ecosphere*, 7 (2016): e01213

——, 'Tracking of Marine Predators to Protect Southern Ocean Ecosystems', *Nature*, 580 (2020): 87–92

Hückstädt, L. A., et al., 'Projected Shifts in the Foraging Habitat of Crabeater Seals Along the Antarctic Peninsula', *Nature Climate Change*, 10 (2020): 472–7

Intergovernmental Panel on Climate Change (IPCC), *The Ocean and Cryosphere in a Changing Climate: Special Report of the Intergovernmental Panel on Climate Change*, 1st edition, Cambridge: Cambridge University Press, 2022

Johnson, C. M., et al., *Protecting Blue Corridors: Challenges and Solutions for Migratory Whales Navigating National and International Seas*, World Wildlife Fund, 2022

Khatiwala, S., et al., 'Global Ocean Storage of Anthropogenic Carbon', *Biogeosciences*, 10, no. 4 (2013): 2169–91

Kim, B.-M., et al., 'Antarctic Blackfin Icefish Genome Reveals Adaptations to Extreme Environments', *Nature Ecology and Evolution*, 3 (2019): 469–78

Laborie, J., et al., 'Estimation of Total Population Size of Southern

Elephant Seals (*Mirounga leonina*) on Kerguelen and Crozet Archipelagos Using Very High-Resolution Satellite Imagery', *Frontiers of Marine Science*, 10 (2023): 1149100

LaRue, M., et al., 'Insights from the First Global Population Estimate of Weddell Seals in Antarctica', *Science Advances*, 7, no. 39 (2021): eabh3674

Leaper, R., and Miller, C., 'Management of Antarctic Baleen Whales Amid Past Exploitation, Current Threats and Complex Marine Ecosystems', *Antarctic Science*, 23, no. 6 (2011): 503–29

Li, Q., et al., 'Abyssal Ocean Overturning Slowdown and Warming Driven by Antarctic Meltwater', *Nature*, 615 (2023): 841–7

Lyne, M., 'Deep Ocean Currents Around Antarctica Headed for Collapse, Study Finds', University of New South Wales, 31 March 2023, www.sciencedaily.com/releases/2023/03/230330102327.htm (accessed 15 May 2024)

Matsuoka, K., and Hakamada, T., 'Estimates of Abundance and Abundance Trend of the Blue, Fin and Southern Right Whales in the Antarctic Areas IIIE–VIW, South of 60°s, Based on JARPA and JARPAII Sighting Data', 2014

Meyer, B., ''The Overwintering of Antarctic krill, *Euphausia superba*, from an Ecophysiological Perspective', *Polar Biology*, 35, no. 1 (2012): 15–37

Moran, A. L., and Woods, H. A., 'Why Might They Be Giants? Towards an Understanding of Polar Gigantism', *Journal of Experimental Biology*, 215, no. 12 (2012): 1995–2002

Nicol, S., and Robertson, G., 'Ecological Consequences of Southern Ocean Harvesting', in N. Gales (ed.), *Marine Mammals: Fisheries, Tourism and Management Issues*, Collingwood, Australia: CSIRO Publishing, 2003, pp. 48–61

Pallin, L. J., et al., 'A Surplus No More? Variation in Krill Availability Impacts Reproductive Rates of Antarctic Baleen Whales', *Global Change Biology*, 29, no. 8 (2023): 2108–21

Rankin, J. C., and Tuurala, H., 'Gills of Antarctic Fish', *Comparative Biochemistry and Physiology Part A: Molecular & Integrative Physiology*, 119 (1998): 149–63

Rocha Jr, R. C., Clapham, P. J., and Ivashchenko, Y., 'Emptying the Oceans: A Summary of Industrial Whaling Catches in the 20th Century', *Marine Fisheries Review*, 76, no. 4 (2015): 37–48

Rowlands, E., et al., 'The Effects of Combined Ocean Acidification and Nanoplastic Exposures on the Embryonic Development of Antarctic Krill', *Frontiers of Marine Science*, 8 (2021): 709763

Savoca, M. S., et al., 'Baleen Whale Prey Consumption Based on High-Resolution Foraging Measurements', *Nature*, 599 (2021): 85–90

Shishido, C. M., et al., 'Polar Gigantism and the Oxygen–Temperature Hypothesis: A Test of Upper Thermal Limits to Body Size in Antarctic Pycnogonids', *Proceedings of the Royal Society B*, 286, no. 1900 (2019): 20190124

Sidell, B. D., and O'Brien, K. M., 'When Bad Things Happen to Good Fish: The Loss of Hemoglobin and Myoglobin Expression in Antarctic Icefishes', *Journal of Experimental Biology*, 209, no. 10 (2006): 1791–802

Siegel, V., 'Krill Stocks in High Latitudes of the Antarctic Lazarev Sea: Seasonal and Interannual Variation in Distribution, Abundance and Demography', *Polar Biology*, 35, no. 8 (2012): 1151–77

Tarling, G., et al., 'Growth and Shrinkage in Antarctic Krill *Euphausia superba* is Sex-Dependent', *Marine Ecology Progress Series*, 547 (2016): 61–78

In a Single Human Generation

Grémillet, D., et al., 'Radar Detectors Carried by Cape Gannets Reveal Surprisingly Few Fishing Vessel Encounters', *PLOS ONE*, 14, no. 2 (2019): e0210328

Sherley, R. B., et al., 'The Status and Conservation of the Cape Gannet *Morus capensis*', *Ostrich*, 90, no. 4 (2019): 335–46

Tew Kai, E., et al., 'Are Cape Gannets Dependent upon Fishery Waste? A Multi-scale Analysis Using Seabird GPS-Tracking, Hydro-Acoustic Surveys of Pelagic Fish and Vessel Monitoring Systems', *Journal of Applied Ecology*, 50, no. 3 (2013): 659–70

INDEX

Page numbers in italics denote illustrations within the text.

INDEX

INDEX

INDEX

INDEX

Toby Strong © Open Planet Studios/Silverback Films

SIR DAVID ATTENBOROUGH is a broadcaster and naturalist whose television career is now in its seventh decade. Over that time he has established himself as the world's leading natural history programme maker with multiple landmark series to his name. His recent book, *A Life on Our Planet*, co-authored with Jonnie Hughes, was an international bestseller.

COLIN BUTFIELD is co-founder and director of Open Planet Studios. He has worked on numerous documentaries and short films including the BBC series *Earthshot*, Netflix features *A Life on Our Planet* and *Breaking Boundaries*, and the National Geographic feature documentary *Ocean*, also with David Attenborough. He is co-author of *Earthshot: How to Save Our Planet*.